Industrial Lasers
and Their Applications

Industrial Lasers
and Their Applications
Second Edition

James T. Luxon
David E. Parker

GMI Engineering & Management Institute
Flint, Michigan

Prentice Hall, Englewood Cliffs, New Jersey 07632

Library of Congress Cataloging-in-Publication Data

Luxon, James T.
    Industrial lasers and their applications / James T. Luxon, David
E. Parker.—2nd ed.
       p. cm.
    Includes bibliographical references and index.
    ISBN 0-13-463803-4
    1. Lasers—Industrial applications. I. Parker, David E.,
II. Title.
    TA1677.L89   1992   91-16689
    621.36'6—dc20   CIP

Editorial/production supervision and interior design: *Harriet Tellem*
Cover design: *Patricia McGowan*
Prepress buyer: *Mary Elizabeth McCartney*
Manufacturing buyer: *Susan Brunke*
Acquisitions editor: *Karen Gettman*
Editorial assistant: *Connie Uccelli*

 © 1992 by Prentice-Hall, Inc.
A Simon & Schuster Company
Englewood Cliffs, New Jersey 07632

The publisher offers discounts on this book when ordered
in bulk quantities. For more information, write:
    Special Sales/Professional Marketing
    Prentice-Hall, Inc.
    Professional & Technical Reference Division
    Englewood Cliffs, New Jersey 07632

Printed in the United States of America

10 9 8 7 6 5 4 3 2 1

ISBN 0-13-463803-4

Prentice-Hall International (UK) Limited, *London*
Prentice-Hall of Australia Pty. Limited, *Sydney*
Prentice-Hall Canada Inc., *Toronto*
Prentice-Hall Hispanoamericana, S.A., *Mexico*
Prentice-Hall of India Private Limited, *New Delhi*
Prentice-Hall of Japan, Inc., *Tokyo*
Simon & Schuster Asia Pte. Ltd., *Singapore*
Editora Prentice-Hall do Brasil, Ltda., *Rio de Janeiro*

To Our Wives,
Sally and Nancy

Contents

Preface to Second Edition

The purpose of this book, as described in the Preface to the first edition, has not changed. In the second edition we have tried to clarify some areas, simplify where necessary, add appropriate material, delete inappropriate or irrelevant material, and update to the extent that it is practical in the rapidly developing field of industrial laser applications. There has been no attempt to make this book an exhaustive encyclopedia of lasers and laser applications. For example, microelectronics applications are not covered; that would require a large book in its own right. What we have done is try to produce an introduction to lasers and industrial applications that provides the fundamentals in a readable fashion and enough information for a student, scientist, or engineer to intelligently and appropriately apply lasers in industrial situations.

Chapter 1, Principles of Optics, has been expanded to include sections on enhanced and antireflection coatings, electro-optical modulators, and optical materials. The section on optical resonators has been moved from Chapter 6, Laser Theory, to Chapter 7, Laser Beam Properties, to provide better continuity in the area of laser optical properties. The notation in Chapter 7 has been significantly simplified, as has the entire discussion of beam propagation and focusing, without sacrificing content or the usefulness of the material. A section on excimer lasers has been added to Chapter 8, Types of Lasers.

Chapters 9 and 10, which deal with low-power applications of lasers, have been expanded to three chapters and a substantial amount of new material has been added. Chapter 9, Using Low-Power Lasers for Alignment, Gaging, and Inspection, deals with devices and techniques used in industry that require structured laser light, but not the accuracy that can be obtained by using interferometry. Similar applications have been grouped and preceded by a discussion of relevant optical design concepts. This is followed by a rather complete discussion of interferometric applications in Chapters 10 and 11. Chapter 10, Introduction to Interferometry, is designed to review fundamentals used in both chapters and discuss industrial applications of traditional interferometric techniques that have benefited by the advent of the laser. Chapter 11, Holographic and Speckle Interferometry, has been expanded and includes new material on thermoplastic holographic plates, carrier fringe techniques, and speckle interferometry.

Chapter 11, Interaction of High-Power Laser Beams with Materials, is Chapter 12 in the new edition. Material that was found to be extraneous or irrelevant has been removed from this chapter, making it much more readable. A new section on material removal has been added, and the section on welding has been expanded. Chapter 12, High-Power Laser Applications, is Chapter 13, High-Power Systems and Applications, in the new edition. A section on beam delivery for high-power systems and a subsection on polarization effects have been added to this chapter.

Many other changes have been made that we believe will increase the value and usefulness of this book. These include corrections, new photographs, and additional problems in many chapters.

James T. Luxon
David E. Parker

Preface to First Edition

The purpose of this book is to provide the reader with an introduction to lasers and their industrial applications. To facilitate this objective, such devices as photodetectors and modulators, which are frequently found in laser applications, are also covered. And to make the book as self-contained as possible, the concepts of basic optics that are pertinent to lasers and their applications are presented. Many engineering students do not cover this material formally in their course work; moreover, many working engineers and scientists either have not had training in optics or have been away from it for a long time and may need a refresher. Some laser theory is presented to provide a working understanding of the laser and to clear away the mysticism surrounding the device. When tools are understood, they are used more frequently and used properly.

The topic of laser beam optics, including propagation, focusing, and depth of focus, is covered in some detail for both Gaussian and higher-order mode beams because such information is of practical value to industrial applications of lasers.

A chapter on optical detectors, including detector arrays, is preceded by a short chapter on semiconductors, to enhance the understanding of solid-state optical devices, and by a chapter in which radiometry, photometry, and optical device parameters are discussed.

It would be impossible to present an exhaustive treatment of the interaction of high-power laser beams and matter, but some of the most pertinent cases are presented in a chapter on laser beam materials interaction. Separate chapters are devoted to industrial applications of low- and high-power lasers. Specific types of applications are presented, along with additional theoretical or conceptual material where required.

This is not a book on lasers but rather a book that is intended to help prepare engineering students or practicing engineers and scientists for the practical application of lasers in an industrial manufacturing setting. Thus the book may be used for a one-semester, junior-senior level course on lasers and laser applications or by practicing engineers and scientists who need to learn quickly the essentials of lasers and their applications. Greater depth in the topics on lasers or materials interactions, for example, can then be obtained from many more advanced books.

This book can be used in several different ways. The first chapter on basic optics can be omitted if the reader has a background in optics. The chapters on semiconductors, parameters, radiometry, and devices can be omitted if the reader's interests do not lie in these areas; these topics are not essential to the remainder of the book.

Readers who have some familiarity with lasers and their properties can omit the overview chapter on lasers. Any chapters on low- or high-power applications may be omitted without loss of continuity.

The authors are greatly indebted to a number of people. We want to express our appreciation to our families for their patience and encouragement. We would like to thank many of our students who gave us constructive criticism and other assistance. We would particularly like to thank Mark Sparschu and Jim McKinley for reading Chapters 9 and 10 and working the problems. We also want to thank Ms. Barbara Parker for her skillful proofreading and Ms. Judy Wing for her patience, skill, and good humor in typing much of the manuscript.

James T. Luxon
David E. Parker

Chapter 1

Principles of Optics

This chapter is intended to provide the reader with a basic working knowledge of the principles of optics, including a description of the nature of electromagnetic radiation, as well as geometrical and physical optics. This chapter also provides a basis for much of what follows in subsequent chapters.

1-1 NATURE OF ELECTROMAGNETIC RADIATION

Electromagnetic radiation exhibits both wavelike and particlelike characteristics, as does matter when it comes in small enough packages, like electrons. Both aspects of electromagnetic radiation are discussed and both are relevant to understanding lasers. From the point of view of its wavelike characteristics, electromagnetic radiation is known to exhibit wavelengths from less than 10^{-13} m to over 10^{15} m. Included in this range, in order of increasing wavelength, are gamma rays, x rays, ultraviolet waves (uv), visible light, infrared (ir) light, microwaves, radio waves, and power transmission waves. Figure 1-1 illustrates the various parts of the range of electromagnetic waves as a function of wavelength.

The sources of gamma rays (γ rays) are nuclear transitions involving

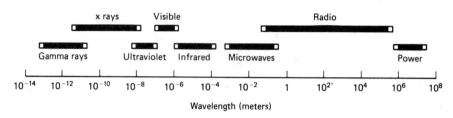

Figure 1-1 Electromagnetic radiation as a function of wavelength.

radioactive decay. X rays are produced through electronic transitions deep in the electronic structure of the atom. Ultraviolet waves result from electronic transitions involving fairly high energy and overlap the x-ray region somewhat. Visible radiation extends from about 0.35 to 0.75 μm and is due to electronic transitions, primarily of valence electrons. Infrared radiation results from electronic transitions at the near visible end and molecular vibrations toward the long-wavelength end. Microwaves and radio waves are produced by various types of electronic oscillators and antennas, respectively.

The term *light* is used loosely to refer to radiation from uv through ir.

The wavelike properties of electromagnetic radiation can be deduced from the wave equation presented here in one-dimensional form:

$$\frac{\partial^2 E}{\partial z^2} = \frac{1}{c^2}\frac{\partial^2 E}{\partial t^2} \tag{1-1}$$

where c is the velocity of light and E is the electric field intensity. The wave equation can be derived from Maxwell's equations, the foundation of all classical electromagnetic theory. The symbol E in Eq. (1-1) may represent any one of the various electromagnetic field quantities, but for our purposes the electric field intensity is of greatest interest.

Another relationship that can be deduced from Maxwell's equations that is of use to us is Poynting's theorem:

$$\mathbf{S} = \mathbf{E} \times \mathbf{H} \tag{1-2}$$

where **S** is power flow per unit area, **E** is electric field intensity, and **H** is magnetic field intensity. For a freely propagating electromagnetic wave, it reduces to

$$S_{\text{ave}} = \tfrac{1}{2}EH \tag{1-3}$$

where S_{ave} is the average power flow per unit area and E and H are amplitudes.

Light may be thought of as being composed of sinusoidal components of electric and magnetic fields from the point of view that electromagnetic radiation is a wave. For a simple electromagnetic wave propagating in an unbounded medium (the electric field varying parallel to a single direction, referred to as *linear polarization*), the wave may be schematically represented as in Fig. 1-2.

The electric and magnetic fields are oriented at right angles to each other and to the direction of propagation z. **E, H,** and z form a right-hand triad; that is, **E** $\times$ **H** gives the direction of propagation. Ordinary (unpolarized) light contains a mixture of polarizations in all directions perpendicular to the direction of propagation. Because of the vector nature of the electric field, unpolarized light can be thought of as an equal mix of electric field strength in orthogonal directions, say x and y, with random phase relations between the various contributions to the electric field. The significance of this statement will become apparent later on.

The speed of propagation in free space (vacuum) is approximately 3×10^8 m/s and is equal to $1/\sqrt{\mu_0\epsilon_0}$ according to classical electromagnetic wave theory. For an electromagnetic wave propagating in a dielectric medium, the speed is

$$\frac{1}{\sqrt{\mu\epsilon}} = \frac{1}{\sqrt{\mu_r\mu_0\epsilon_r\epsilon_0}} = \frac{c}{\sqrt{\mu_r\epsilon_r}}$$

where c is the speed of light in free space and μ_r and ϵ_r are the relative permeability and permittivity of the medium, respectively. In nonmagnetic materials $v = c/\sqrt{\epsilon_r}$. The refractive index of a dielectric medium is defined by

$$n = \frac{c}{v} \tag{1-4}$$

and so it is seen that $n = \sqrt{\epsilon_r}$ for most dielectrics.

It is possible to show that $E = ZH$, where Z is called the intrinsic impedance of the medium. This fact can be used to put Eq. 1-3 in the form

$$I = \frac{1}{2}\frac{E^2}{Z} \tag{1-5}$$

where I is the irradiance (power per unit area). You may recognize the similarity between Eq. (1-5) and the equation for Joule heating in a resistor,

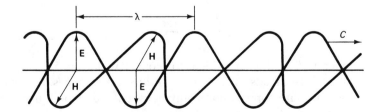

Figure 1-2   Propagation of a linearly-polarized electromagnetic wave.

which is $P = V^2/R$, where P is power, V is voltage, and R is resistance. The one-half does not appear in the Joule's law heating equation because V is a root mean square rather than an amplitude.

The intrinsic impedance of an unbounded dielectric is $Z = \sqrt{\mu/\epsilon}$, where $\sqrt{\mu_0/\epsilon_0}$, the intrinsic impedance of free space, is 377 Ω. Then

$$Z = \frac{377 \ \Omega}{\sqrt{\epsilon_r}} = \frac{377 \ \Omega}{n} \qquad (1\text{-}6)$$

for nonmagnetic dielectrics. Equation (1-5) can therefore be written

$$I = \frac{1}{2} \frac{E^2 n}{377 \ \Omega} \qquad (1\text{-}7)$$

The subject of electromagnetic wave propagation in conductors is beyond the scope of this book, but a few pertinent facts can be pointed out. The impedance of a good conductor is given by

$$Z = \sqrt{\frac{\mu\omega}{\sigma}} \ e^{-j(\pi/4)} \qquad (1\text{-}8)$$

where ω is the radian frequency of the light and σ is the conductivity of the conductor. As can be seen from Eq. (1-8), Z is a complex impedance. E is very small in a conductor; H is large. When an electromagnetic wave strikes a conductor, E will go nearly to zero and H becomes large due to large induced surface currents. The results are considerable reflectance of the incident wave and rapid attenuation of the transmitted wave. The skin depth, which is a measure of how far the wave penetrates, is given by

$$\delta = \frac{1}{\sqrt{\pi\mu\sigma f}} \qquad (1\text{-}9)$$

For frequencies of interest in this book, chiefly visible and ir to 10.6 μm, δ is extremely small and absorption can be assumed to take place at the surface for all practical purposes.

Based on the Drude free electron theory of metals, it can be shown that the fraction of the incident power absorbed by a metal is given approximately by

$$A = 4\sqrt{\frac{\pi c \epsilon_0}{\lambda \sigma}} \qquad (1\text{-}10)$$

The reflectance is $R = 1 - A$, which, for copper with $\sigma = 5.8 \times 10^7$ (Ω-m)$^{-1}$ at $\lambda = 10.6$ μm, leads to $R = 0.985$. Actual reflectances may exceed this value for very pure copper. For highly polished or diamond-turned copper mirrors, the reflectance exceeds 0.99.

The particlelike behavior of electromagnetic radiation is exhibited in many experiments, such as the photoelectric effect and Compton scattering. It was in an explanation of the photoelectric effect in 1905 that Einstein proposed that electromagnetic radiation (light for short) is composed of bundles of energy, quanta, which are referred to as *photons*. The energy of each of these photons, he argued, is hf, where h is Planck's constant ($h = 6.6 \times 10^{-34}$ J·s). Planck had determined this constant previously in explaining the dependence of black-body radiation on frequency.

In 1924, DeBroglie proposed a mathematical model for the photon. This model consists of an infinite sum of waves of different frequencies within a finite frequency range with an appropriate amplitude function. It was really nothing more than a Fourier integral representation of a finite pulse, with the amplitude function chosen to produce a wave packet with minimum uncertainty products. A schematic representation of such a pulse is given in Fig. 1-3. The outer solid lines form an envelope of the amplitude of the actual wave. The uncertainties referred to concern the length of the packet L, its relation to the uncertainty in its momentum Δp, and the relation between frequency bandwidth and the time it takes the photon to pass a given point. These relations are

$$\Delta pL \geq \frac{h}{2\pi}$$

$$\Delta f\, \Delta t \geq \frac{1}{2\pi} \tag{1-11}$$

These relations will be useful later in discussing the coherence of light sources.

DeBroglie proposed a duality of both light and matter; that is, he suggested that matter should exhibit both wave and particle characteristics. It was later shown that electrons can be diffracted by crystals, and the observed wavelength agreed with that predicted by DeBroglie. The DeBroglie wavelength can be deduced by setting Einstein's famous mass–energy relationship $E = mc^2$ equal to the energy of a photon hf. Thus

$$E = mc^2 = hf$$

and $mc = hf/c$ is the momentum of a photon; therefore, momentum and

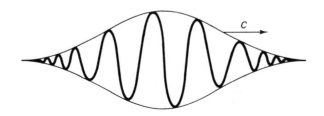

Figure 1-3   Schematic representation of mathematical model of a photon.

wavelength are related by

$$p = mc = \frac{h}{\lambda} \tag{1-12}$$

The relation holds for both light and matter waves. These results can be used to show (left as an exercise for the student) that $dp/dt = P/c$ for a total absorption at a surface, where P is total power in the beam and c is the speed of light. The radiation pressure is dp/dt divided by the area of incidence. In general, this pressure can be written

$$\frac{F}{A} = \frac{(1 + R)P}{cA} \tag{1-13}$$

where R is the fraction of the incident power reflected and A is the area of incidence. This pressure can be substantial for a focused laser beam.

1-2 REFLECTION AND REFRACTION

The *law of reflection* states that, for specular reflection of light, the angle of incidence equals the angle of reflection. This situation is illustrated in Fig. 1-4. The reflected ray lies in the same plane as the incident ray and the normal to the surface. This plane is referred to as the *plane of incidence*. A specular surface is one with a surface finish characterized by rms variations in height and separation of peaks and valleys (surface roughness) much less than the wavelength of the light. In other words, a surface that is not a good specular surface in the visible could be quite specular at longer wavelengths. This is an important point to remember when working around high power (long-wavelength) lasers.

Most surfaces cause reflected light to contain a portion of specular and diffusely reflected (scattered) light. The diffuse reflection is the result of random reflections in all directions due to roughness of surface finish. An ideal diffuser scatters equal amounts of power per unit area per unit solid angle in all directions. Hence a perfect diffusing surface is equally bright from all viewing angles. Few surfaces approach the ideal case.

The law of refraction, or *Snell's law* as it is commonly called, is given by Eq. (1-14) and the angles are defined in Fig. 1-5.

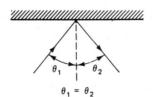

$\theta_1 = \theta_2$ **Figure 1-4**   Law of reflection.

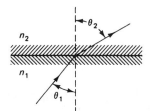

Figure 1-5 Refraction at a dielectric interface.

$$n_1 \sin \theta_1 = n_2 \sin \theta_2 \tag{1-14}$$

The law of reflection applies equally to all materials, whereas Snell's law, in the form given in Eq. (1-14), is valid only for an interface between two dielectrics.

Two phenomena of importance relate simply to Snell's law. The first, total internal reflection (TIR), occurs when light travels from a medium of higher refractive index into one of the lower refractive index. Because $n_2 < n_1$, Snell's law requires that $\theta_2 > \theta_1$. At an angle of incidence called the critical angle, θ_2 becomes 90°. At this angle of incidence θ_c, and for all $\theta_1 > \theta_c$, all incident power is reflected. The critical angle is deduced from Snell's law to be

$$\theta_c = \sin^{-1} \frac{n_2}{n_1} \tag{1-15}$$

The second phenomenon is Brewster's law, which has significance in many laser designs as well as other areas. Brewster's law states that, when the reflected and refracted rays are at right angles to each other, the reflected light is linearly polarized perpendicular to the plane of incidence. The angle of incidence at which this occurs is called the Brewster angle θ_B. Figure 1-6 illustrates this phenomenon. Because θ_2 is the complement of θ_B, Snell's law gives

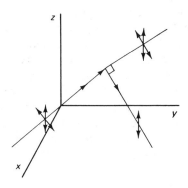

Figure 1-6 Illustration of Brewster's law. Interface is the yz plane and rays are parallel to the xy plane.

$$n_1 \sin \theta_B = n_2 \cos \theta_B$$

or   (1-16)

$$\theta_B = \tan^{-1}\left(\frac{n_2}{n_1}\right)$$

This is a reversible phenomenon, unlike TIR, and the Brewster angle from side 2 to 1 is $90° - \theta_B$ or $\tan^{-1}(n_1/n_2)$.

When light is incident on a surface, a certain fraction of the light is absorbed or transmitted and the remainder is reflected. The fraction of power reflected, called the reflectance R, for light normally incident on an interface between two dielectrics is given by

$$R = \left(\frac{n_1 - n_2}{n_1 + n_2}\right)^2$$ (1-17)

If light is incident from side 1, there will be a 180° phase shift in the reflected wave for $n_2 > n_1$, but no phase shift if $n_2 < n_1$. This factor is important for antireflection and enhanced reflection coatings on optical components.

For nonnormal angles of incidence, the Fresnel formulas provide the reflectances for parallel and perpendicular polarizations.

$$R_{\parallel} = \left[\frac{\tan(\theta_1 - \theta_2)}{\tan(\theta_2 + \theta_1)}\right]^2$$

$$R_{\perp} = \left[\frac{\sin(\theta_2 - \theta_1)}{\sin(\theta_1 + \theta_2)}\right]^2$$ (1-18)

Note that $R_{\parallel}$ goes to zero for $\theta_2 + \theta_1 = 90°$, the Brewster angle condition.

1-3 MIRRORS AND LENSES

In this section the results derive from the application of the law of reflection to mirrors and Snell's law to thin and thick lenses. Pertinent information concerning spherical aberration is also presented.

The sign convention used here, with regard to lenses and mirrors, follows that of Jenkins and White, 1976. In this convention, light is always assumed to be traveling left to right. All object or image distances measured to the left of a reflecting surface are positive; otherwise, they are negative. Object distances to the left and image distances to the right are positive for refracting surfaces; otherwise, they are negative. Radii of curvature are positive if measured in the direction of the reflected or refracted light; otherwise, they are negative. Lenses or mirrors that converge parallel rays have positive focal lengths and negative focal lengths if they diverge parallel rays. An object height measured above the axis is positive; below the axis it is negative.

Figure 1-7 illustrates the effect of concave (positive) and convex (negative) spherical mirrors on parallel rays.

Parallel rays are reflected through a point called the focal point F for the concave mirror. Point C locates the center of curvature, and F lies midway between C and the vertex of the mirror. Hence f, the focal length or distance from F to the vertex, equals $r/2$, where r is the radius of curvature of the mirror. For the convex mirror, rays parallel to the axis are reflected such that they appear to be coming from the focal point F. For the concave mirror, rays parallel to the axis are reflected through the focal point. In both cases, the rays are reversible. These facts can be used to construct images graphically for a given object. Graphical image construction is depicted in Fig. 1-8. Analytically, object distance o, image distance i, and f are related by

$$\frac{1}{o} + \frac{1}{i} = \frac{1}{f} \tag{1-19}$$

where all distances are measured from the mirror vertex. Also, it is easy to see that magnification m is given by

$$m = \frac{Y_i}{Y_o} = -\frac{i}{o} \tag{1-20}$$

Equations (1-19) and (1-20) apply equally well to spherical mirrors and thin lenses if the sign convention is strictly followed. Actually, not all parallel

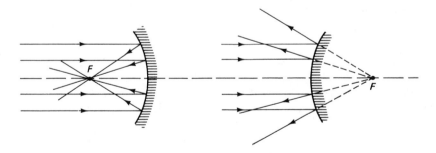

Figure 1-7   Effect of concave and convex mirrors on rays parallel to the axis.

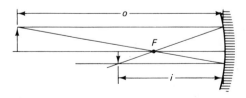

Figure 1-8   Image construction for a spherical mirror.

rays are focused at a common point. This fact leads to what are referred to as *aberrations*, the most important of which, for our purposes, is spherical aberration. Spherical aberration occurs because rays parallel to the axis are focused closer to the mirror vertex the farther they are from the axis. Also, parallel rays not parallel to the axis suffer an aberration called *astigmatism* (nothing to do with astigmatism of the eye). Strictly speaking, a spherical mirror is good only for paraxial rays, that is, rays parallel and close to the axis. Parabolic mirrors do not suffer from spherical aberration; all rays parallel to the axis are focused at a common point, but astigmatism is more severe.

Spherical mirrors are little used for precisely focusing purposes except in large telescopes where the focusing is on axis and the spherical aberration can be corrected. Spherical mirrors are used in lasers as resonator mirrors. Here spherical aberration is not a problem because the wavefronts produced are spherical, not plane as we have been tacitly assuming.

Mirrors used for off-axis focusing, such as for high-power lasers, are usually nonspherical or aspheric. This condition is accomplished by using a sector of a parabolic surface as illustrated in Fig. 1-9.

Many other types of mirrors are used with lasers, such as axicons, so called because they are specifically designed to produce a particular axial distribution of light on reflection (axicons may also be transmissive and are used for spherical aberration correction). A variation, the waxicon (W-shaped axicon), is used to convert the ring-mode output of an unstable resonator to a near-Gaussian distribution. Such a mirror is depicted in Fig. 1-10. Another type of mirror focuses the ring mode to a thin ring of finite radius, as shown in Fig. 1-11.

Thin lenses can be treated similarly to mirrors except that the analytical results are derived from Snell's law instead of the law of reflection. The thickness of the lens is neglected and two focal points are required. The primary focal point F_p is defined such that light rays coming from it, or headed toward it, are refracted parallel to the axis. The secondary focal point F_s is defined such that rays traveling parallel to the axis are refracted so that they pass through F_s or appear to be coming from it. Figure 1-12 illustrates these comments about focal points for converging and diverging lenses. Note that the refraction is assumed to take place at a single plane

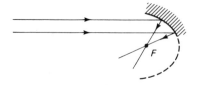

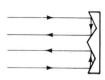

Figure 1-9 Parabolic off-axis focusing mirror.

Figure 1-10 Waxicon.

Figure 1-11   Axicon.

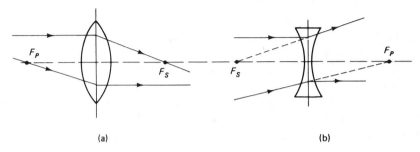

(a)   (b)

Figure 1-12   Primary (F_p) and secondary (F_s) focal point for (a) converging and (b) diverging thin lenses.

referred to as a *principal plane*. The focal lengths are also measured from this plane. If the refractive index of the lens is n' and the refractive indices of the surrounding media are $n = n'' = 1$, then

$$\frac{1}{o} + \frac{1}{i} = \frac{1}{f} = (n' - 1)\left(\frac{1}{r_1} - \frac{1}{r_2}\right) \tag{1-21}$$

where r_1 and r_2 are the radii of curvatures of the first and second surfaces, respectively. The magnification is the same as for the mirrors.

Lenses with spherical surfaces suffer from the same aberrations as mirrors, plus several others, including chromatic aberration. Different wavelengths have different refractive indices; therefore, different wavelengths are focused at different points. In the context of this book, the monochromatic aberrations, spherical aberration, coma, and astigmatism are the important ones. *Coma*, like astigmatism, occurs for parallel rays that are not paraxial and causes a lateral smearing of the image. Coma and astigmatism are minimized by keeping the beam as nearly parallel to the axis as possible. Curvature of field and distortion are important in imaging applications.[1] The correction for spherical aberration will be discussed after thick lenses are covered.

In many practical applications, such as focusing of high-power lasers for materials processing, the thin-lens assumption is not valid. Not everything previously developed need be discarded, however. In fact, by defi-

[1] See, for example, *Fundamentals of Optics* by Jenkins and White.

nition of two principal planes, properly located, the thin-lens equation remains valid. The primary focal length is measured from the primary principal plane, and the secondary focal length is measured from the secondary principal plane. The principal planes and their relation to the focal lengths and the bending of rays are illustrated in Fig. 1-13.

The lens equation becomes

$$\frac{n}{o} + \frac{n''}{i} = \frac{n}{f} = \frac{n''}{f''} \tag{1-22}$$

which is, in fact, the same equation as for a thin lens. It was assumed before that $n = n''$, but it will be instructive to assume $n \neq n''$ for awhile. Note in Fig. 1-13 where the focal lengths are measured from and where the rays appear to bend.

To determine the focal lengths of a thick lens, the focal lengths for refraction at a single surface must be known. The single-surface focal length equations are given by Eq. (1-23) and relate to Fig. 1-14.

$$\frac{n}{f_1} = \frac{n'}{f_1'} = \frac{n' - n}{r_1} \qquad \text{first surface}$$

$$\frac{n_1'}{f_2'} = \frac{n_1''}{f_2''} = \frac{n'' - n'}{r_2} \qquad \text{second surface} \tag{1-23}$$

Equation (1-24) gives the relationship between the thick-lens primary and secondary focal lengths in terms of the single-surface focal lengths and the appropriate refractive indexes.

$$\frac{n}{f} = \frac{n''}{f''} = \frac{n'}{f_1'} + \frac{n''}{f_2''} - \frac{dn''}{f_1'f_2''} \tag{1-24}$$

Equations (1-25) give the distances of the principal planes (H, H'') from the lens vertices, V_1 and V_2, as defined in Fig. 1-13.

$$V_2H = (d - f)\frac{d}{f_2'}, \qquad V_2H'' = -f''\frac{d}{f_1'} \tag{1-25}$$

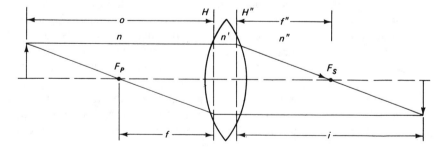

Figure 1-13   Thick lens with principal planes H and H''.

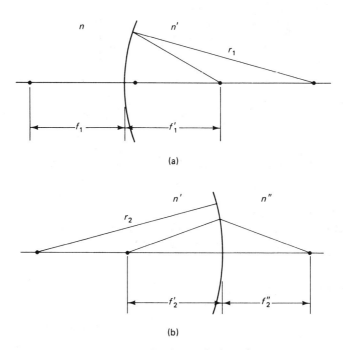

Figure 1-14   Refraction at single surfaces.

The procedure for correcting for spherical aberration in spherical lenses is, in effect, to increase the curvature of both refracting surfaces away from the direction of the oncoming light. An optimum correction exists for all focal lengths and refractive indices, but spherical aberration cannot be eliminated for a single-element lens. Figure 1-15 depicts a partially corrected plano-convex lens and a fully optimized meniscus lens. The light must enter the lens from the most curved side; otherwise, spherical aberration is increased. Sherman and Frazier have developed a method for comparing plano-convex and meniscus lens to determine if the superior correction of the meniscus lens warrants its higher cost. The spherical-aberration-limited spot diameter d_{sa} (blur of an ideal focused spot) is given by Eq. (1-26).

$$d_{sa} = k \frac{D^3}{f^2} \tag{1-26}$$

Figure 1-15   (a) Partially corrected plano-convex, (b) optimally corrected meniscus lens for 10.6-μm wavelength.

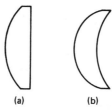

TABLE 1-1   Spherical Aberration Factor k for
Selected Materials at 10.6 μm

Material	Meniscus	Plano-convex
Ge	0.0087	0.0295
ZnSe	0.0187	0.0286
GaAs	0.0114	0.0289

The factor k depends on whether the lens is a plano-convex or meniscus, and the refractive index of the lens, D, is the diameter of the beam entering the lens containing at least 95% of the beam power. Some values of k are listed in Table 1-1 for $\lambda = 10.6$ μm.

Methods for calculating the diffraction-limited spot size, with which d_{sa} should be compared, are discussed later.

1-4 BEAM EXPANDERS

Because of their importance in laser applications, some discussion of beam expanders is presented here.

Lens beam expanders can be made with two positive or one negative and one positive lens. These types of transmissive beam expanders, along with two reflective beam expanders, are depicted in Fig. 1-16. In either case, the lenses have a common focal point. By similar triangles, it is seen from

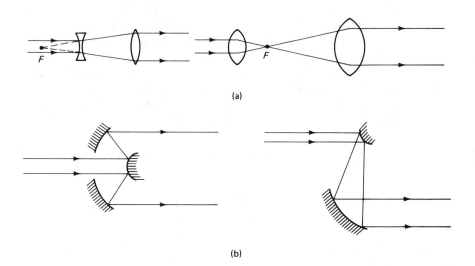

(a)

(b)

Figure 1-16   Beam expanders: (a) transmissive, (b) reflective.

Fig. 1-16 that

$$W_2 = \frac{f_2}{f_1} W_1 \qquad (1\text{-}27)$$

for either lens combination if the absolute values of the focal length of the negative lens is assumed. It is important to note that the diffraction spreading of the beams is reduced in proportion to f_1/f_2 because of the increase in effective aperture.

   The nonfocusing beam expander is practical for most applications since it can be made more compact. For very high power applications, it is undesirable to focus the beam in air due to the possibility of breakdown in air or distortion of the beam due to induced refractive index changes. The focusing beam expander is used when beam filtering is required. This process basically involves focusing the beam on a pinhole and filtering (blocking) out the higher-order diffraction fringes. The beam that exits the pinhole contains much less spatial noise and hence has a more uniform irradiance distribution.

1-5 INTERFERENCE AND DIFFRACTION

A brief discussion of interference and diffraction phenomena is presented in this section, particularly as they relate to lasers and laser applications.

   When two or more electromagnetic waves having some fixed-phase relationship relative to one another are superimposed, the result is referred to as *interference*. The electric field intensities must be added rather than simply adding the powers or irradiances, as is done when phase relationships are timewise random, such as with adding the light from two flashlights. If the beam from a laser is split into two beams and then these separate beams are recombined on a screen, as depicted in Fig. 1-17, the phase relation between the beams will vary from point to point on the screen due to slight

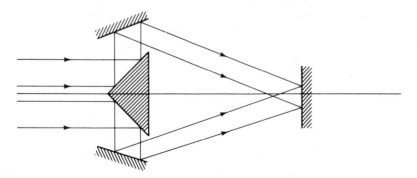

Figure 1-17   Split laser beam recombined on screen.

path-length differences. This situation can be expressed mathematically as in Eq. (1-28):

$$E_1 + E_2 = E_{10} \sin(2\pi f t) + E_{20} \sin(2\pi f t + \theta) \qquad (1\text{-}28)$$

where θ is a phase angle that will vary, depending on location on the screen, and E_{10} and E_{20} are the amplitudes. Since the resultant amplitude is of primary interest, the electric fields can be treated as phasors (vectors in the complex plane with time dependence omitted) with an angle θ between them. Using the law of cosines produces

$$E^2 = E_{10}^2 + E_{20}^2 + 2E_{10}E_{20} \cos \theta \qquad (1\text{-}29)$$

Since irradiance is proportional to the electric field amplitude squared, Eq. (1-29) can be written

$$I = I_1 + I_2 + 2\sqrt{I_1 I_2} \cos \theta \qquad (1\text{-}30)$$

It must be remembered that irradiance is a time-averaged quantity, and if θ varies randomly with time, the last term in Eq. (1-30) goes to zero, leaving the usual result for addition of noise signals. When θ does not vary randomly with time, an interference pattern is produced on the screen with a visibility V defined by

$$V = \frac{I_{\max} - I_{\min}}{I_{\max} + I_{\min}} \qquad (1\text{-}31)$$

It is left as an exercise to show that Eq. (1-31) reduces to $2I_1 I_2 / (I_1 + I_2)$. Clearly, $0 \le V \le 1$.

There are numerous ways to produce interference effects with light. Just the propagation of a plane or spherical wave (any shape for that matter) is the result of continuous interference of an infinity of Huygen's wavelets that make up the propagating wavefront. Huygen's principle provides a convenient model for the construction of propagating wavefronts, given the wavefront and its direction of propagation at some place and time. According to Huygen's principle, every point on a wavefront is a source of a spherical wave, with directionally dependent amplitude, called a *Huygen's wavelet*. Each wavelet expands forward (in the direction of propagation of the original wavefront) at a rate equal to the velocity of propagation of the original wavefront. The shape and location of the wavefront after time t are determined by constructing wavelets with radius ct and drawing a wavefront tangent to all possible wavelets. Such a construction is depicted in Fig. 1-18.

When such a wavefront is interrupted by an aperture or an edge, some contributors to the propagating wavefront are removed. The result is a diffraction pattern such as is commonly produced by a knife edge. This situation is schematically represented in Fig. 1-19. This diffraction pattern can be used for precise location of a sharp edge.

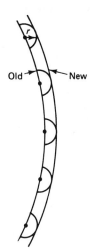

Figure 1-18 Illustration of Huygen's principle.

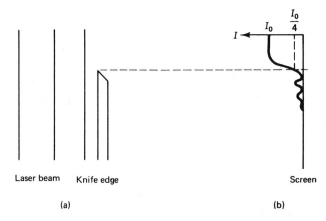

Figure 1-19 Knife-edge experiment with laser beam uniformly illuminating the knife edge.

Let's look at what happens when a laser beam illuminates two thin, rectangular slits in a screen. The experimental setup is depicted in Fig. 1-20. This particular example is referred to as *Young's double-slit interference*. All parallel rays leaving the slits are focused at a common point in the focal plane of the lens. The *optical path lengths* traveled by any two parallel rays starting from a common perpendicular plane are equal. The path-length difference between oblique pairs of rays can be represented as in Fig. 1-20, where a path-length difference of λ is indicated for two parallel rays. These rays interfere constructively in the focal plane, because $\theta = 360°$. If the obliquity is increased until the difference is 2λ or $\theta = 720°$, constructive interference will again be obtained. A general equation can be written.

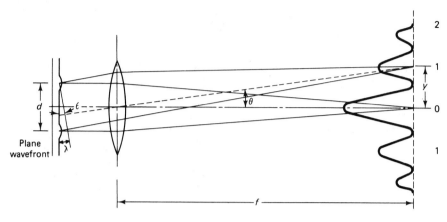

Figure 1-20 Double-slit experiment.

$$m\lambda = d \sin \theta \qquad (1-32)$$

where $m = 0, 1, 2, \ldots$. Note, also, that $y = f \tan \theta$.

If hundreds or thousands of slits or rulings are provided instead of just two, it is easy to show that Eq. (1-32) still applies, and we have what is called a *diffraction* (interference) *grating*. The number of rulings (lines) per centimeter is usually given, and it equals the reciprocal of d. Because of the multiplicity of lines, nearly total destructive interference occurs for values of θ only slightly different than those given by Eq. (1-32). Hence a diffraction grating can be used to resolve (separate) closely spaced wavelengths in a beam of light. Diffraction gratings are frequently used as a means of separating laser lines and in wavelength selection, or tuning, in lasers.

Another very important diffraction phenomenon is the effect of an aperture on a beam of light. Figure 1-21 depicts a laser beam incident on a rectangular slit. In this case, it is not easy to locate the points of constructive interference; so we look for the points of destructive interference, which are of more interest anyhow. The slit is divided into two equal parts of width $D/2$, and a ray is taken from the top of each part. If these rays are parallel and differ in path by $\lambda/2$ or 180°, then they will interfere destructively as shown. All the rest of the rays, parallel to these two, will also interfere destructively in pairs. To find higher-order minima, simply divide the slit into 4, 6, . . . equal parts, and let the path-length difference between rays from the tops of adjacent parts be $\lambda/2$. A general equation can be developed (left as an exercise):

$$m\lambda = D \sin \theta \qquad (1-33)$$

where $m = 1, 2, 3, \ldots$. Note that Eq. (1-33) has the same form as Eq. (1-32) but a different interpretation. The width of the image line produced on

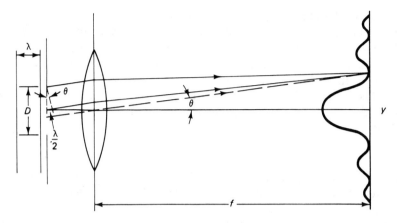

Figure 1-21   Single-slit experiment.

the screen is taken to be the distance between first minima, $2y$. For small angles,

$$\lambda = D\theta = \frac{Dy}{f} \tag{1-34}$$

An extremely important case of diffraction is that caused by a circular aperture. The derivation for this case is beyond the scope of this book. The result for the first minimum is

$$\sin \theta = \frac{1.22\lambda}{D} \tag{1-35}$$

Any beam of light made up of *plane waves* truncated by a circular aperture must have a minimum divergence angle given by Eq. (1-35), where D is the truncated diameter. Equation (1-35) is frequently applied to laser beams, letting D be the diameter of the output beam. This is, at best, only a rough estimate because laser beams are not accurately represented by a uniform plane wave.

1-6 OPTICAL COATINGS

Optical coatings are used on lenses, windows, laser rods, and other optical elements to eliminate unwanted reflections and on dielectric mirrors to enhance reflection. More complex coatings may be used than will be discussed here, but the principle of operation is the same. All such coatings function on the basis of thin-film interference.

Antireflection (AR) coatings, in their simplest form, are quarter-wavelength coatings with a refractive index intermediate to the substrate and air.

In this case the phase shift for reflected light at each interface is identical and can be ignored. For normal incidence the path-length difference is a half-wavelength; this produces a 180° phase shift as shown in Fig. 1-22. Consequently, the light reflected from the first surface interferes destructively with the light reflected from the second surface.

This is not sufficient to eliminate reflection entirely. The reflectance from each surface must also be equal. Equating the two reflectance values leads to

$$n_c = \sqrt{n_s} \tag{1-36}$$

During vacuum coating, the refractive index can be adjusted to achieve the result to a high degree of accuracy. To make the coating more rugged, the thickness can be made an odd number of quarter-wavelengths. However, coating absorption increases, which is a serious problem in high-power laser applications.

Enhanced reflection coatings in their simplest form are composed of

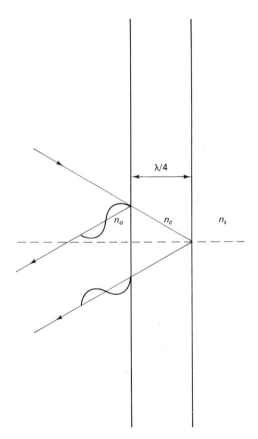

Figure 1-22 AR coating.

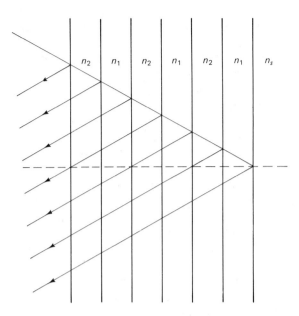

Figure 1-23 ER coating.

many quarter-wavelength layers of alternating refractive index such that the phase change between successive layers is always 180°, so the net phase shift for reflected light between successive surfaces is 360°. The result is that constructive interference occurs for light reflected from all surfaces. Figure 1-23 illustrates this process.

1-7 COHERENCE

Coherence of electromagnetic waves is a complex subject, and no attempt at a rigorous treatment will be made here. The general nature of coherence is relatively simply, however, and a useful working knowledge can be presented fairly easily.

Essentially, two waves are coherent if there is some fixed-phase relationship between them. Two pure sinusoidal waves of the same frequency would have perfect coherence. Unfortunately, physical phenomena can, at best, only be approximated by pure sinusoids. In reality, there are two manifestations of coherence, spatial and temporal.

Spatial coherence refers to correlation in phase at the same time but at different points in space. For the experiments on interference and diffraction previously described, a high degree of spatial coherence was necessary for interference or diffraction patterns to occur. Spatial coherence can be produced with ordinary sources in two ways. One is to place the source far enough away from the slits, diffraction grating, and so forth, so that each emitting point of the source uniformly illuminates the entire device. The second is to place a small aperture (pinhole) in front of the source, such as a gas discharge lamp. Because of the small emitting area of the source viewed by the diffracting device, substantial spatial coherence is achieved. Lasers emit radiation with nearly perfect spatial coherence across the entire beam.

Temporal coherence refers to correlation in phase at the same point in space at different times. As a wave passes through a given point in space, its phase will undergo random changes. These changes are unpredictable as to when they will occur or as to what the change in phase will be. The origin of these changes is at the source. The average time between changes in phase for an isolated atom is the lifetime of the atom in its excited state, which averages about 10^{-8} s. The average emission time for luminous gases, liquids, or solids, is much shorter due to line-broadening mechanisms. Remember, $\Delta f \tau \simeq 1$ according to the uncertainty relations. For lasers operating in a single mode,[2] the coherence time τ_{coh} can be much longer, approaching that and, in many cases, bettering that of the isolated atom (molecule).

The coherence time is simply the approximate time that it takes a photon to pass a given point in space. Thus the coherence length L_{coh}, which is the approximate length of the photon, is related to the coherence time by

$$L_{coh} = c\tau_{coh} \tag{1-37}$$

The Michelson–Morley experiment exemplifies the significance of temporal coherence. This experiment, which uses the Michelson interferometer, is depicted in Fig 1-24. Let's assume a laser beam is used as the light source. The beam is split into two equal parts by the beam splitter. Waves 1 and 2 are reflected by the corresponding mirrors and are partially recombined at the beam splitter and observed at the top. For interference to be observed over a long period of time, such as required by the eye, a photographic emulsion, or most photodetectors, waves 1 and 2 must overlap. Otherwise, there will be no fixed-phase relationship between waves numbered 1 and those numbered 2. Thus

$$2(L_1 - L_2) > L_{coh}$$

and there will be no observable interference. This setup has been used to measure coherence length or photon length for decades. Before the advent of the laser, the practical application of the Michelson interferometer was limited to situations where $L_1 \simeq L_2$. Lasers may have coherence lengths easily in excess of 100 m; therefore, one leg, the reference leg, can be a few centimeters long, whereas the other leg may be nearly 50 m long. Appli-

---

[2] Lasers may operate in a variety of axial and transverse electromagnetic (TEM) modes. They are discussed later.

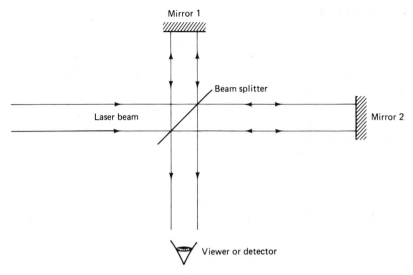

Figure 1-24   Michelson interferometer.

cations of the Michelson interferometer and related interferometric techniques, such as holography, will be discussed later.

Some further insight into the concept of coherence and how it can be measured is obtained from the following simplified approach. A beam may be thought of as consisting of a completely coherent part CI and a completely incoherent part $(1 - C)I$. Because there is no correlation between these parts, they can be added like noise signals:

$$I = CI + (1 - C)I$$

where $0 \le C \le 1$. If two such beams are superimposed, the coherent parts will add like phasors (coherently):

$$I_{coh} = (CI) + (CI) + 2\sqrt{CICI} \cos \theta$$

whereas the incoherent parts will add like noise signals:

$$I_{incoh} = (1 - C)I + (1 - C)I = 2(1 - C)I$$

The total irradiance is obtained by simply adding these last two expressions:

$$I_T = I_{coh} + I_{incoh} \qquad (1\text{-}38)$$

The quantity C could represent the fraction of overlap in a Michelson–Morley experiment or it could represent the degree of spatial coherence in a double-slit or other diffraction experiment. It is left as an exercise to show that Eq. (1-38) leads to $V = C$ from the definition of visibility in Eq. (1-30). Thus the fraction of overlap or degree of spatial coherence is easily measured.

1-8 POLARIZATION

The subject of linear polarization has, of necessity, already been dealt with. Because electromagnetic radiation is a vector phenomenon, however, other more general forms of polarization can be obtained from the principle of superposition. Also, some discussion of how different forms of polarized light may be produced is in order.

If you imagine superimposing two electromagnetic waves linearly polarized at right angles to one another but with a 90° phase difference between them, the result is circular polarization. Figure 1-25 is a series of head-on views of these waves; each point is advanced one-eighth of a wavelength. The tip of the resultant vector is performing uniform circular motion in the plane of E_1 and E_2 and has an amplitude $E_r = E_1 = E_2$. In this view, E_2 lags E_1 by 90°, and so left-hand rotation occurs; if the phase relation is reversed, right-hand rotation occurs. The actual path of the tip of the resultant vector is a spiral in the direction of propagation. If $E_1 \neq E_2$, then elliptical polarization occurs with the major axis lying parallel to the larger E field. If the phase angle is not 90°, elliptical polarization occurs with the major axis lying at some angle between E_1 and E_2. Figure 1-26 depicts elliptically polarized light from a head-on view. The phase angle between E_1 and E_2 is given by

$$\sin \theta = \frac{E_b}{E_a} \qquad (1\text{-}39)$$

regardless of the relative amplitudes of E_1 and E_2. E_a is the y-axis intercept, E_b is the maximum projection on the y-axis.

Linearly polarized light is produced in many ways. It occurs in gas lasers when Brewster windows are used to seal the ends of the gas tube or in solid lasers if the ends of the rod are cut at the Brewster angle. Linear polarization also arises when repeated reflections occur from mirrors, for

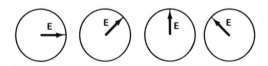

Figure 1-25   Head-on view of circularly polarized light in $\frac{1}{8}\lambda$ increments, left-hand rotation.

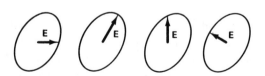

Figure 1-26   Head-on view of elliptically polarized light.

the reflectance is always higher for $E_\perp$ than it is for $E_\parallel$. Lasers with 45° mirrors to fold the beam through a number of tubes have this characteristic. Repeated reflection from a stack of plates oriented at the Brewster angle can linearly polarize an ordinary beam. Polaroid is used extensively for polarizing light, and it consists of molecules that preferentially absorb light polarized parallel to their long axis but not their short axis. A sheet of Polaroid is produced by orienting these molecules with all their long axes parallel. The polarizing direction of the material is at 90° to the long axis of the molecules.

Figure 1-27 illustrates an important relationship between polarizing direction and irradiance called *Malus's law*. Linearly polarized light is incident on a polarizer at an angle θ relative to the polarizing direction. If the incident amplitude is E_0, the transmitted amplitude is $E_0 \cos \theta$. Therefore, the transmitted irradiance is related to the incident irradiance by

$$I = I_0 \cos^2 \theta \qquad (1\text{-}40)$$

Several other types of devices produce linearly polarized light from ordinary light more efficiently and with less attenuation than Polaroid, such as Nicole and Glan prisms, which are based on the optical properties of transparent solids. Space does not permit discussion of all these polarizers. Most standard optics books describe them in detail.

It is important to discuss briefly some techniques for producing circular and elliptically polarized light. Many materials exhibit *birefringence*, which means two different refractive indices for the perpendicular components of light passing through the material. Such materials have an optical axis, and

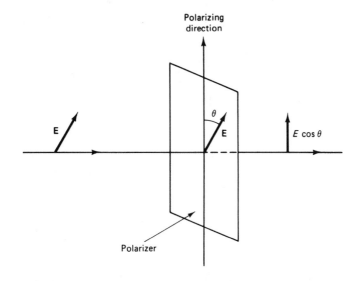

Figure 1-27   Illustration of Malus's law.

a unique refractive index occurs for a ray with the electric field vector perpendicular to the optical axis—the normal ray. The refractive index for the other ray, the extraordinary ray, varies from the value for the ordinary ray, depending on the material, up to a maximum or minimum for the electric field vector parallel to the optic axis. One such material is the crystal calcite, which is the basis for the Nicole prism. A ray traveling parallel to the optic axis in calcite travels faster than one perpendicular to it. Because of the differing refractive indices, the two polarized beams can be separated. The property of birefringence is also exhibited by some crystallized polymers and can be induced in a variety of crystals and liquids by application of a strong external electric field. Examples are potassium dihydrogen phosphate (KDP) and lithium niobate crystals and nitrobenzene, a liquid. They are referred to as Pockels cells and Kerr cells, respectively. In both cases, high voltages are required, but the optical axis can be switched on and off rapidly, in nanoseconds. Besides, the amount of birefringence (refractive index difference) is a function of applied voltage. Hence the phase difference between the two polarizations, after passing through the medium, can be controlled by the applied voltage. Pockels cells will be covered in Section 1-11.

Let's consider a device that alters the phase between two rays by 90°; such devices are referred to as quarter-wave plates. They could be a sheet of birefringent polymeric material, a Pockels cell, or a Kerr cell. Linearly polarized light enters the device such that the direction of polarization makes a 45° angle with the optic axis. Resolving the electric field into components parallel and perpendicular to the optic axis results in two beams traveling at different speeds. This situation is depicted in Fig. 1-28. Normally, a fast axis and a slow axis are specified in the material. The beam with polarization parallel to the fast axis advances just one quarter-wavelength, 90°, relative

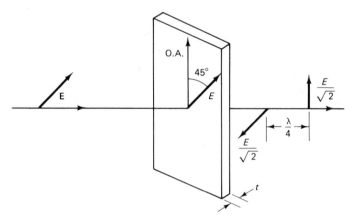

Figure 1-28 Quarter waveplate. O.A. is the optical axis and t is appropriate thickness to provide 90° phase shift ($\lambda/4$) between the orthogonal components.

to the other beam on passing through the material. Hence the beam exiting the quarter-wave plate is circularly polarized. If E is not 45° to the optic axis, then elliptically polarized light is produced. If the thickness corresponds to some phase-angle difference other than 90°, elliptical polarization occurs. A phase change of $\lambda/2$ or 180° simply rotates the direction of polarization by 90°. These are important factors in many laser applications, such as rapid Q-switching of lasers and conversion of a linearly polarized laser beam to circularly polarized to improve cutting ability.

An interesting consideration is that of the superposition of two counterrotating circularly polarized light beams. If the beams have equal amplitudes, the result will be linearly polarized light with twice the amplitude of one of them. Figure 1-29 will aid in explaining this situation. As the electric field vectors rotate, their sum remains parallel to the bisector of the angle between them. If the phase of one of the beams is changed, however, say by moving a mirror from which it is reflected, the orientation of the resultant **E** vector will be changed. A $\lambda/8$ motion of the mirror causes a $\lambda/4$, 90° phase change, which produces a 45° rotation of the resultant **E** vector. Not only can this be used to determine how far the mirror has moved, but the direction of rotation of the resultant **E** vector indicates which way the mirror has moved.

A final phenomenon of particular interest when it is essential that light reflected from optical components not be fed back into the laser is the Faraday effect. Some materials, such as lead glass, become *optically active* when subjected to a strong magnetic field parallel to the direction of propagation. Optical activity means that linearly polarized light entering this material will have its polarization direction rotated. The amount of rotation is given by

$$\phi = VB1 \tag{1-41}$$

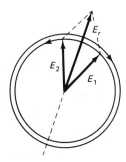

Figure 1-29   Superimposed counterrotating, circularly polarized beams. (Drawn at slightly different amplitudes for clarity.) E_r is the resultant and is linearly polarized with amplitude equal to $2E_1$ for $E_1 = E_2$.

where B is the magnetic field intensity, 1 is the length of material, and V, the Verdet constant, has a value of 110 $(\text{tesla·meter})^{-1}$ for dense flint glass.

If linearly polarized light is passed through a Faraday rotator that rotates the polarization direction 45°, any light that is reflected back through the rotator will be rotated an additional 45°, for a total of 90°. This light, traveling in the reverse direction back toward the laser, is easily attenuated by a polarizer oriented to transmit the original light beam.

1-9 FIBER OPTICS

The subject of fiber optics is included here because of the growing use of fiber optics in optical instrumentation, communication, and laser beam delivery for surgical and other important applications. Optical fibers may be either glass or plastic. When low losses are important, glass fibers are superior, having losses of less than 1 dB/km.[3]

The operation of simple optical fibers is based on the phenomenon of total internal reflection. If a fiber core is clad with a material of lower refractive index than the core, there will always be a critical angle such that rays striking the interface of the core and cladding at an angle greater than this angle will be totally reflected. Thus the fiber becomes a light guide or dielectric waveguide. This phenomenon is illustrated in Fig. 1-30.

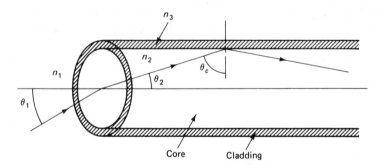

Figure 1-30 Optical fiber showing entering ray striking cladding at critical angle.

In Fig. 1-30 a ray is shown entering the face of the fiber at the maximum possible angle that will result in total internal reflection of the ray inside the fiber. Applying Snell's law at the fiber face and assuming the ray strikes the core–cladding interface at the critical angle, we can show that

$$\text{NA} = n_1 \sin \theta_1 = (n_2^2 - n_3^2)^{1/2} \tag{1-42}$$

Here n_1, n_2, and n_3 are the refractive indices of the external medium, core,

---

[3] The loss in decibels is defined as $10 \log_{10} P_{in}/P_{out}$, where P_{in} and P_{out} are the input and output optical power, respectively.

and cladding, respectively, θ is the half-angle of the entrance cone, and NA stands for numerical aperture.

The cladding in glass fibers is also glass, but doped differently than the core to give it a slightly lower refractive index. For example, if the core and cladding indices are 1.55 and 1.50, respectively, for a fiber in air, NA = 0.39 and $\theta_1 = 23°$.

The type of fiber just discussed is a step index fiber. Another type commonly used is the graded index fiber. Here the refractive index decreases radially from the center in a parabolic fashion. Rays launched at small angles to the axis of the fiber continuously change direction toward the axis without ever striking the cladding. The path is a zigzag combination of straight-line segments between reflections in a step index fiber, whereas the path in a graded index fiber is essentially sinusoidal.

Losses in glass fibers are primarily due to scattering by impurities and defects. Losses in plastic fibers are chiefly due to absorption.

Short delay times in optical communications are desirable to minimize pulse spreading and the resultant degradation of information. There are two causes of delay in optical fibers, referred to as modal dispersion and wavelength dispersion. Modal delay is caused by the different distances traveled by different rays within the fiber. The maximum value for this type of delay in a step index fiber can be deduced by considering a ray traveling parallel to the axis and a ray reflecting from the cladding at the critical angle. The time delay per unit of fiber length is then given by

$$\Delta t = \frac{n_2}{c} \left(\frac{1}{\sin \theta_c} - 1 \right) = \frac{n_2}{c} \left(\frac{n_2}{n_3} - 1 \right) \tag{1-43}$$

This amounts to 0.17 ns/m for the fiber example discussed. Actual delay times are less because the losses are higher for the rays traveling greater distances.

Optical fibers are waveguides and, as such, not every imaginable geometrical path for a ray is an allowable mode. In fact, if the fiber is made sufficiently thin, typically a few micrometers in diameter, only one allowed mode exists. Such single-mode fibers effectively eliminate modal delay, and picosecond delay times are possible. Graded index fibers minimize modal dispersion because the farther a ray wanders from the axis the more time it spends in a region of lower refractive index (higher speed). Thus the time delay between axial and nonaxial rays is reduced.

Wavelength dispersion, or delay due to the variation of refractive index with wavelength, is minimized by proper material design to reduce this variation and by the use of narrow-bandwidth light emitters, chiefly diode lasers.

It must be recognized that losses also occur due to reflection at entrance and exit surfaces and coupling of light from emitters into fibers and the coupling of light out of fibers to detectors. Highly efficient techniques and devices have been developed, however, for coupling light in and out of fibers and for coupling light from one fiber to one or several other fibers.

1-10 ACOUSTO-OPTICAL MODULATORS

Acousto-optical modulators are used for the production of short, high-power pulses from lasers, beam deflection, amplitude modulation, and frequency modulation. The variety of laser applications makes a discussion of such devices pertinent.

    An acousto-optical modulator is a device that uses a transparent solid block of material, such as fused quartz, to which is attached a piezoelectric transducer. Acoustic waves are produced in the quartz by applying a high-frequency voltage, say 40 MHz with 0.5-V amplitude, to the transducer. The vibration of the piezoelectric transducer sends acoustic waves through the quartz block. This process is depicted in Fig. 1-31. A laser beam is passed through the block at a small angle relative to the acoustic wavefronts. Light reflected from the wavefronts can interfere constructively only if Eq. (1-44) is satisfied:

$$m\lambda_0 = 2\lambda_a \sin \theta, \qquad m = 1, 2, 3, \ldots \qquad (1\text{-}44)$$

where λ_0 is the optical wavelength and λ_a is the acoustic wavelength. With care, only the first-order, $m = 1$, deflection occurs along with the original beam. Equation (1-44) may be recognized as similar to Bragg's law for x-ray or electron diffraction from crystals, where λ_a would be the interplanar spacing.

    The amplitude of the diffracted beam can be varied from zero to a maximum by varying the voltage applied to the piezoelectric transducer.

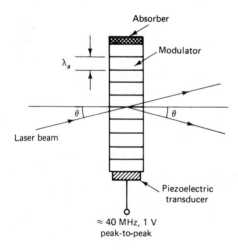

Figure 1-31   Acousto-optical modulator.

The frequency of the diffracted beam is shifted as a result of the Doppler effect. According to relativity theory, when an observer and a light source have a relative velocity v, the observed frequency is given by

$$f = f_o \frac{1 + (v/c)}{\sqrt{1 - (v^2/c^2)}} \tag{1-45}$$

where f_o is the frequency if observer and source are at rest relative to each other. In this case, $v \ll c$ and so Eq. (1-45) may be written

$$\frac{\Delta f}{f_o} = \frac{v}{c} \tag{1-46}$$

The velocity is positive for relative motion toward each other and negative for motion away from each other. If the light is reflected from a moving mirror, the Doppler shift is doubled. In the acousto-optical modulator

$$\frac{\Delta f}{f} = \frac{2v_a \sin \theta}{c} \tag{1-47}$$

because the component of acoustic velocity in the beam direction is $v_a \sin \theta$. Combining Eqs. (1-47) and (1-44) for $m = 1$ gives

$$\Delta f = \pm f_s \tag{1-48}$$

The plus sign is taken if the acoustic wave velocity component is opposite the beam propagation direction, and the minus sign is taken if the acoustic velocity has a component in the direction of beam propagation.

Acousto-optical modulators can be used for deflection (usually less than 0.5°) with over 1000 resolvable spots, amplitude modulation, and frequency modulation.

1-11 ELECTRO-OPTICAL MODULATORS

Electro-optical modulators are used intracavity for Q-switching of lasers for short pulses in the nano- to microsecond range. They are also used external to the cavity for switching and phase alteration to change the polarization state of light.

The electro-optical effect is based on the phenomenon of birefringence induced in a material by an applied electric field. The change in refractive index due to an applied electric field is given by

$$\Delta \left(\frac{1}{n^2} \right) = rE + PE^2 \tag{1-49}$$

where r is the coefficient for the linear electro-optical effect (Pockels) and P is the coefficient for the quadratic electro-optical effect (Kerr). Only the Pockels effect is of interest here.

TABLE 1-2 Pockels Cell Materials

Material	$r \, (10^{-12} \, \text{m/V})$
Potassium dihydrogen phosphate (KDP)	10.6
Deuterated potassium dihydrogen phosphate (KD*P)	26.4
Lithium tantalate	30.3
Lithium niobate	30.8

The materials used for Pockels-effect devices, called Pockels cells, are crystalline and may be naturally birefringent, such as lithium tantalate (LiTaO$_3$), or naturally isotropic, like gallium arsenide (GaAs). Table 1-2 contains a list of a few materials suitable for Pockels cell, along with their linear-optic coefficients.

Both longitudinal and transverse applications of the electric field are utilized. We will consider only the more common longitudinal application here. A Pockels cell is schematically represented in Fig. 1-32.

From Eq. (1-49), for small changes in refractive index

$$\Delta n \approx n - n_o = \pm \tfrac{1}{2} r n_0^3 E \qquad (1\text{-}50)$$

The plus sign gives the refractive index change relative to one of the preferred directions (parallel or perpendicular to the optical axis), and the minus sign gives the change for the other axis. For the sake of discussion, assume that

$$n_x - n_o = \frac{1}{2} r n_0^3 E$$

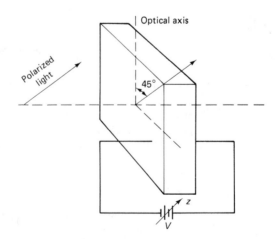

Figure 1-32 Longitudinal Pockels cell.

and

$$n_y - n_o = -\frac{1}{2} rn_0^3 E$$

Then

$$n_x - n_y = rn_0^3 = \frac{rn_0^3 \, V}{l}$$

where l is the thickness of the cell.

The change in phase caused by this difference in refractive index (remember, the waves are traveling at different speeds) is given by

$$\phi = \frac{2\pi}{\lambda} (n_x - n_y)1 = \frac{2\pi}{\lambda} rn_0^3 V \qquad (1\text{-}51)$$

Assuming the crystal is properly oriented and the light entering the cell is linearly polarized at 45° to the optical axis, an appropriate voltage can be applied to convert the light to circularly polarized light (90° phase shift) or rotate the direction of polarization 90° (180° phase shift). The latter case is used for switching, since a polarizer placed after the cell oriented with its polarizing direction parallel to the incoming light's direction of polarization will absorb the transmitted light when the voltage is applied. This arrangement can also be used to modulate the amplitude of a light beam.

Certain materials, KDP for example, may also be used to alter the frequency of light. This is a second-order effect and requires an intense beam of light. From a photon point of view, it is a matter of two or more photons being absorbed and a new one being emitted. From conservation of energy, the new photon's frequency is the sum of the energies of the incident photons. Different frequencies may be mixed, but our interest here is simply frequency doubling. In this case, an intense laser beam enters a crystal and emerges with twice the frequency of the incident beam. Since momentum must also be conserved, it is necessary that the refractive indices for the two different frequencies be identical (this is called *phase matching*). This can be achieved in birefringent crystals for certain specific directions of propagation relative to the optical axis.

1-12 OPTICAL MATERIALS FOR LASERS

The choice of material used either internal or external to the laser depends on its function, the wavelength at which the laser operates, and the power produced by it. For lasers that operate in the visible or near infrared ($\lambda = 0.35$ to 2.0 μm), both transmissive and reflective optics are nearly always made from BK-7, a borosilicate-crown glass. This material is very stable, has low absorption, and has a refractive index of 1.505 at a wavelength of 1.06 μm.

For some ultraviolet laser applications and mid-infrared lasers, metallic or semiconductor substrates are used for reflective optics. Copper is popular for both internal and external applications due to its high thermal conductivity and the ease with which it can be machined. Many copper mirrors are diamond turned for mid-infrared applications.

For internal applications in the mid-infrared spectral region, when extreme thermal conditions exist, silicon is superior to copper because it has a very low thermal coefficient of expansion. It is also transparent at a wavelength of 10.6 μm, so some power can be leaked through it for power measurement. Reflectance is controlled by a silver or gold coating. Germanium is sometimes used for this purpose because it has better transmittance than silicon in the infrared.

External reflective optics are generally copper with a variety of possible coatings, polished copper, or bare diamond-turned surfaces. Copper surfaces are very easily damaged, even by lens tissues. Molybdenum (moly) serves as an extremely rugged coating, but it has an absorption of 2.0% to 3.0% compared with less than 1.0% for bare copper. Silver coatings can provide as low as 0.3% absorption. Gold is also used and is more rugged than copper or silver, but it has slightly higher absorption than silver. Table 1-3 contains the absorptivities (percentage of power absorbed) for bare copper and various coatings.

Transmissive optics for the mid-infrared spectral region are generally semiconductor materials, although the salts sodium chloride and potassium chloride have high transmittance in this region as well. Unfortunately, the salts are hygroscopic and defy application of AR or ER coatings. Germanium is suitable for applications up to about 100 W. Beyond that zinc selenide (ZnSe) is the most commonly used material. Gallium arsenide (GaAs) has high transmittance, but unlike ZnSe it is not transparent in the visible part of the spectrum.

Because it has lower absorptivity and refractive index change with temperature, ZnSe is preferable in relatively high transmittance applications

TABLE 1-3 Mirror Absorptivities

Material	Absorptivity (%)
Unprotected copper	1.4
Molybdenum	2.0–3.0
Unprotected gold	1.1
Protected silver	0.9
Enhanced silver	0.5
Superenhanced silver	0.3

TABLE 1-4 Optical Properties of Selected Materials

Material	n	Absorptivity (cm^{-1})	Thermal Expansion (10^{-6}/°C)	Refractive Index Change (10^{-6}/°C)	Thermal Conductivity (W/cm °C)
GaAs	3.27	<0.01	5.7	149	0.47
Ge	4.0	<0.0018	5.7	277	0.59
ZnSe	2.4	<0.0005	7.6	60	0.043

(50% or higher). GaAs is preferable for high-reflectance applications (95% or higher) because it has a higher thermal conductivity and lower coefficient of thermal expansion.

A problem that sometimes occurs with transmissive optics used for output couplers, windows, or lenses for multikilowatt CO_2 lasers is thermal lensing or focusing. This phenomenon exists for two reasons. Absorption of light by the optical element either in the bulk or the ER or AR coatings causes a nonuniform temperature rise, with the maximum temperature at the center, or hot spot, of the beam. Thermal expansion of the element causes it to bulge, thus producing a positive lensing effect. An order of magnitude larger effect is caused by the variation of refractive index with temperature, which is positive for semiconductor optical materials. This causes an effective path-length increase in addition to that caused by thermal expansion. Since the thermal coefficient of refractive index change is about eight to twenty-five times larger than the coefficient of thermal expansion, the refractive index gradient produced by the temperature gradient is the major contributor to thermal lensing. Permanent deformation of the optical element may occur due to the development of internal stresses.

If the primary absorption is in the coating, as it is with ER coatings, a high-thermal-conductivity material is substrate desirable. If the major absorption is in the substrate material (as is the case with AR coatings), then a low-absorptivity material is best. Table 1-4 contains a listing of selected optical material properties for use at a wavelength of 10.6 μm.

PROBLEMS

1-1. Show that

$$E = E_0 \sin(\omega t - kx) \quad \text{and} \quad E = E_0 e^{j(\omega t - kx)} \quad (j = \sqrt{-1})$$

are solutions of Eq. (1-1) and determine the relationship between c, ω, and k.

1-2. A laser beam of 2.5-kW average power is focused to a spot of 50-μm diameter in a medium whose refractive index is 1.5. Determine the irradiance and the electric field amplitude.

1-3. Copper has an electrical conductivity of 6×10^7 $(ohm \cdot m)^{-1}$. Calculate the magnitude of copper's intrinsic impedance at the frequency of a HeNe laser ($\lambda = 0.6328$ μm).

1-4. Calculate the skin depth for the data given in Problem 1-3.

1-5. Calculate the theoretical reflectance for highly polished copper for Nd–YAG laser radiation ($\lambda = 1.06$ μm).

1-6. (a) Calculate the momentum of a photon with a wavelength of 10 μm (CO_2 laser).
   (b) Calculate the pressure exerted by a 1-terrawatt beam focused to a spot of 150-μm diameter.

1-7. Calculate the Brewster angle and the critical angle for light traveling from a medium of 1.5 refractive index (glass) into a medium of 1.33 refractive index (water).

1-8. Brewster windows are to be attached to the ends of a gas laser tube. The windows have a refractive index of 1.5 and are to be oriented so that any light reflected from them is totally polarized perpendicular to the plane of incidence (hence the name). At what angle relative to the tube axis should these windows be oriented?

1-9. Calculate the pertinent reflectances for light incident normally and at 35° to the normal for light traveling in air incident on a medium of 1.4 refractive index.

1-10. An object is placed on the axis 50 cm from the vertex of a concave spherical mirror that has a radius of curvature of 30 cm. Locate the image and determine the magnification. Illustrate with a ray diagram.

1-11. Repeat Problem 1-10 for a convex mirror of the same radius of curvature.

1-12. By careful drawing of a ray diagram, demonstrate spherical aberration for a concave spherical mirror.

1-13. An object is placed on axis 100 cm from a 20-cm focal length positive thin lens. Locate the image and calculate the magnification. Illustrate with a ray diagram.

1-14. Repeat Problem 1-13 for a negative lens of -20-cm focal length.

1-15. A thick meniscus lens of ZnSe used for focusing high-power CO_2 laser beams has a refractive index of 2.4, thickness of 0.92 cm, and radii of curvature of $r_1 = 11.2$ cm and $r_2 = 30$ cm. Locate the principal planes and determine the focal length of the lens (assume that the lens is to be used in air).

1-16. Calculate the spherical-aberration-limited spot size of a CO_2 laser beam ($\lambda = 10.6$ μm) of 2-cm diameter (95% of the beam power falls in this area) focused by a 12.5-cm focal length GaAs lens for both meniscus and plano-convex lenses (see Table 1-1).

1-17. Design two types of beam expanders, using lenses that will expand a 2.5-cm-diameter beam to 10 cm. Do not use a lens with a focal length whose absolute value is less than 10 cm.

1-18. Two coherent laser beams interfere with one another on a screen. The irradiances are 10 and 15 mW/cm². Calculate I_{max}, I_{min}, and V. Calculate I at a phase-angle difference of 30°.

1-19. Assuming that the diffraction spreading (divergence) of a HeNe laser beam (λ = 0.6328 μm) can be approximated by assuming it is equivalent to the diffraction caused by a plane monochromatic wave incident on a circular aperture of diameter equal to the laser beam diameter, calculate the divergence for a 2-mm-diameter beam.

1-20. The frequency bandwidth of a HeNe laser (0.6328 μm) is 100 MHz. Estimate its coherence time and coherence length.

1-21. (a) Calculate the minimum thickness and refractive index for an AR coating on a ZnSe lens whose refractive index is 2.4 for a wavelength of 10.6 μm. (b) What is the next smallest allowable thickness for the coating?

1-22. Sketch a ten-pair ER coating on BK-7 glass that has a refractive index of 1.505 for a wavelength of 1.06 μm. The coatings have refractive indices if 1.4 and 1.6.

1-23. A crossed polarizer is placed in front of a plane-polarized light beam. A second polarizer is placed between the first and the light source such that its polarizing axis makes an angle of 30° with the E field of the incident light. Assuming 10 mW of light power incident on the second polarizer, how much power would be detected after the light has passed through both polarizers?

1-24. Calculate the refractive index difference per centimeter in lithium niobate ($LiNbO_3$) for an applied field of 1000 V/cm. How long must a longitudinal Pockels cell be to cause a phase change of 90° between the fast and slow rays for ruby laser light (λ = 0.6943 μm)?

1-25. Calculate the numerical aperture and half-angle of the entrance cone for light entering a step index optical fiber with core refractive index of 1.55 and cladding index of 1.45. Assume that the fiber is in air.

1-26. Calculate the modal dispersion time delay per kilometer of length for the fiber in Problem 1-25.

1-27. (a) Sketch the layout and polarization orientation for a system to produce modulated polarized light from unpolarized laser light. (b) Calculate the voltage applied to a KD* P Pockels cell to produce a 90° phase shift (n_o for KD* P is 1.52).

1-28. What angle should a HeNe laser beam make relative to the normal, to the surface of an acousto-optical modulator operated at 40 MHz with an acoustical wavelength of 60 μm? What is the shift in frequency of the optical wave, assuming the optical wave has a propagation component in the same direction as the acoustical wave?

REFERENCES

ARNOLD, J. B., R. E. SLADKY, and others, "Machining Nonconventional-shaped Optics," *Optical Engineering,* **16**, No. 4, July–August 1977.

JACKSON, J. D., *Classical Electrodynamics*. New York: John Wiley & Sons, Inc., 1962.

JENKINS, J. A., AND W. E. WHITE, *Fundamentals of Optics*, 4th ed. New York: McGraw-Hill Book Co., 1976.

LACY, E. A., *Fiber Optics*. Englewood Cliffs, NJ: Prentice Hall, 1982.

MANSELL, D. N., AND T. T. SAITO, "Design and Fabrication of a Nonlinear Waxicon," *Optical Engineering,* **16,** No. 4, July–August 1977, 355–59.

MUSIKANT, S., *Optical Materials, An Introduction to Selection and Application.* New York: Marcel Dekker, Inc., 1985.

NUSSBAUM, A., AND R. A. PHILLIPS, *Contemporary Optics for Scientists and Engineers.* Englewood Cliffs, NJ: Prentice–Hall, 1976.

SMITH, W. J., *Modern Optical Engineering.* New York: McGraw-Hill Book Co., 1966.

WILLIAMS, C. S., AND C. A. BECKLUND, *Optics: A Short Course for Engineers and Scientists.* New York: John Wiley & Sons, Inc., 1972.

YARIV, A., *Quantum Electronics*, 3rd. ed. New York: John Wiley & Sons, Inc., 1989.

YOUNG, M., *Optics and Lasers.* New York: Springer-Verlag, 1977.

ZIOCK, K., *Basic Quantum Mechanics.* New York: John Wiley & Sons, Inc., 1969.

Chapter 2

Solid-State Concepts

The purpose of this chapter is to provide the reader who does not already have a strong enough solid-state background with an overview of the concepts required for an understanding of semiconductor optical devices. A short review of some atomic physics concepts is presented, and then the concept of energy bands in crystals is discussed and related to the *pn* junction and transistors. Some simple small-signal amplifier circuits are described.

2-1 ATOMIC STRUCTURE

The Bohr theory of the hydrogen atom (actually any one-electron system), particularly its successor, the Bohr–Sommerfeld theory, was eminently successful in providing a model of electronic structure of the atom. Although this theory is not valid for systems with more than one electron, the scheme of orbitals and suborbitals devised for hydrogen serves conceptually as a starting point for building up the electronic structure of all elements of the periodic table. The quantum mechanics later developed by Schrödinger and Heisenberg proved to be the correct nonrelativistic theory of the atom but, except for a difference in the interpretation of orbital angular momenta, leads to the same results for hydrogen as the Bohr–Sommerfeld model. It is the results of the work of Schrödinger and Heisenberg that are discussed here. The quantum mechanical analysis of the hydrogen atom leads to the conclusion that the electron associated with the nucleus can occupy various energy levels for finite periods of time without emitting radiation. When the electron makes a transition from a higher-energy level E_2 to a lower-energy level E_1, the energy given off as radiation is

$$E_2 - E_1 = hf$$

The various states that the electron can occupy are described by a set of four quantum numbers. Three deal with the three dimensions, or degrees of freedom, required to describe the electron's motion around the nucleus, and the fourth relates to the intrinsic angular momentum of the electron. The intrinsic angular momentum of the electron is a relativistic phenomenon and does not appear in the Schrödinger or Heisenberg formulations, but it does in the later relativistic development due to Dirac. These four quantum numbers are designated by n, ℓ, m_ℓ, and m_s. As implied, the first three are a natural result of the solution of the Schrödinger wave equation for the hydrogen atom, and the fourth, in general, is simply tacked on to account for the relativistic nature of electrons. The principal quantum number n determines the shells, in chemical or spectroscopic terminology, and has integer values from one to infinity. Symbolically, the numbers $n = 1, 2, 3, 4, \ldots$ are represented by $K, L, M, N, \ldots$. The principal quantum number physically relates to the average distance of the electron from the nucleus and its energy.

The quantum number ℓ can take on values $\ell = 0, 1, 2, \ldots, n - 1$ and determines the orbital angular momentum of the electron,

$$L = \sqrt{\ell(\ell + 1)}\, \hbar$$

($\hbar = h/2\pi$) and its energy to some extent. Note that the angular momentum is zero for $\ell = 0$, suggesting that the electron travels in a straight line back and forth through the nucleus (in the Bohr–Sommerfeld theory $\ell = 1, 2, 3, \ldots n$). It presents no conceptual difficulty, for according to DeBroglie, both the electron and the nucleus behave like waves and can therefore pass through one another, like electromagnetic waves or water waves, without necessarily interacting.

The third quantum number, m_ℓ, is the magnetic quantum number and has values of $m_\ell = -\ell, -(\ell - 1) \ldots 0 \ldots (\ell - 1), \ell$. Physically, this quantum number gives the projection of the orbital angular momentum onto an applied (internal or external) magnetic field. L can have projections on a magnetic field of $-\ell\hbar \ldots 0 \ldots \ell\hbar$. The reason L tends to line up with a magnetic field is because of the magnetic moment associated with the orbital angular momentum of the electron. Unlike a classical magnet, the magnetic moment associated with the electronic orbital motion cannot align itself perfectly with the magnetic field, but can take only certain orientations as prescribed by the quantum rules.

The final quantum number, m_s, sometimes called the spin quantum number (even though there is no such classical analog), has values of $\pm\frac{1}{2}$ and physically gives the orientation of the electron's intrinsic angular momentum relative to an applied magnetic field. The allowed projections are $\pm\frac{1}{2}\hbar$. The intrinsic angular momentum of the electron is

$$\left[\frac{1}{2}\left(\frac{1}{2} - 1\right)\right]^{1/2} \hbar = (3/4)^{1/2}\hbar$$

It is easily seen then that the K shell ($n = 1$) has $\ell = 0$, $m_\ell = 0$, and $m_s = \pm\frac{1}{2}$. At this point we invoke the Pauli exclusion principle, which says that *no two electrons can occupy the same quantum state.* In other words, no two electrons can have the same set of quantum numbers if they are in the same system. A system might be a single atom, molecule, or an entire crystal. Therefore, the K shell can have, at most, two electrons in it. Here we presume that this quantum number scheme, derived for hydrogen, will apply to atoms with many electrons, although we do not expect the energy levels to be the same.

Following this same reasoning, the L shell has $\ell = 0$, 1, $m_\ell = -1, 0$, 1, and $m_s = \pm\frac{1}{2}$ for each distinct set of ℓ and m_ℓ, providing for a total of eight electrons. The maximum number of electrons in any shell is $2n^2$.

In spectroscopic terms, the various values of ℓ represent subshells within each shell and are denoted, for historical reasons, by $s, p, d, f, g,$ . . . for $\ell = 0, 1, 2, 3, 4, 5, $ . . . . Thus the electronic structure of sodium can be written

$$1s^2 2s^2 2p^6 3s^1$$

where the numbers represent the principal quantum numbers or shells and the letters represent the ℓ values or subshells. The superscripts are the numbers of electrons in each subshell. Sodium has an atomic number of 11; so 11 electrons should be associated with a neutral sodium atom.

It is the physical nature of atoms, like all natural systems, that they tend toward their lowest energy configuration, which involves achieving full subshells. Sodium can do so either by gaining another electron or by giving one up, as it does in NaCl and in its metallic form. In NaCl the chlorine atom picks up the electron to close its $3p$ subshell, thus forming an ionic bond with sodium. Materials with ionic bonds are insulators.

In such metals as sodium or copper, the valence electrons are mutually shared by atoms in a large region of the material. It is this mutual sharing of electrons that provides the binding forces, the metallic bond.

It is easy to see why sodium and copper are metals: both have unfilled s subshells. It is not so obvious why magnesium, which has closed subshells, is a metal or why silicon, which at a casual glance does not appear to have filled subshells, is an insulator at low temperatures and conducts electricity at room temperature. To explain metals like magnesium and semiconductors like silicon, the energy band structure of crystals is introduced.

2-2 ENERGY BANDS

To develop the notion of energy bands qualitatively, we will return to sodium as an example. Imagine 10^{22} sodium atoms so widely separated that they do not interact. There are 10^{22} electrons occupying 2×10^{22} separate $3s$ states. Now imagine gradually bringing these 10^{22} atoms together to form a single crystal with a volume of about 1 cm^{-3}. As they approach each other within a few atomic radii, the outer energy levels of atoms begin to overlap and become perturbed. The perturbation causes the energy levels to shift, spreading out into a band of 2×10^{22} different, but closely spaced, energy levels. Figure 2-1 is a schematic representation of this thought experiment. Here the atoms start out at $r = \infty$ and end up at $r = r_0$, the equilibrium spacing of the atoms. The original 2×10^{22} $3s$ levels spread out into a band of 2×10^{22} separate levels shared by the entire assemblage of atoms. According to the Pauli exclusion principle, this band will be half-filled by electrons. This means that there are 1×10^{22} unoccupied energy states into which electrons can move to take part in electrical or thermal conduction. Hence sodium is an electrical conductor (metal in this case) and has an electronic contribution to its thermal conductivity.

Note that the p levels in Fig. 2-1 also spread out into an energy band that overlaps the $3s$ band. Thus sodium would have been a conductor even

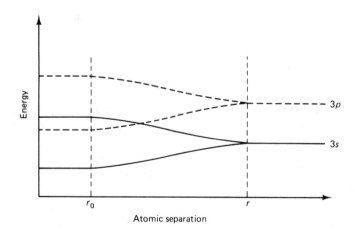

Figure 2-1   Energy levels of Na as a function of atomic separation; r_0 is the equilibrium atomic separation.

if the $3s$ band had been filled, which explains why the elements of the second column of the periodic table, like magnesium, are metals.

Materials like diamond, germanium, silicon, and gallium arsenide are more difficult to explain due to the nature of the covalent bonding in these materials. Covalent bonding is the mutual sharing of electronics between atoms. Consider carbon, which has atomic number 6. We might expect its electronic structure to be $1s^2 2s^2 2p^2$. It turns out that, for purely quantum mechanical reasons that are beyond the scope of this book, the $2s$ and $2p$ levels hybridize into two separate sets of four sp^3 levels.[1] When the crystal is formed, these separate sets of levels form nonoverlapping bands with equal numbers of states in each. Because the total number of hybrid sp^3 states per atom is eight, there will be four states per atom in each of the bands. If the energy gap between the bands is large, the four electrons per atom will go into the lower energy band. In diamond the energy bands are separated by a gap of about 6 eV.[2] Since the average thermal energy at room temperature is about 0.01 eV, very few electrons are able to jump the gap to take part in electrical or thermal conduction. Therefore, diamond is an excellent electrical insulator. (It happens to be an excellent thermal conductor, too, but obviously for a different reason.) When the gap between the separate sp^3 bands is not so great, the result is a semiconductor, such as silicon or germanium. Essentially, the compound semiconductors like gallium arsenide are semiconductors for the same reason.

The distinction between metals, semiconductors, and insulators can be summarized in a simplified energy band scheme as given in Fig. 2-2. A partially filled band is indicative of a metal; a material with a relatively small energy gap will be a semiconductor; a material with a relatively large energy gap is an insulator. Note, however, that at a high enough temperature, assuming they do not melt first, all insulators become semiconductors.

Physically, the energy gap in a semiconductor or insulator represents the energy required to break a covalent bond. When sufficient thermal energy is accumulated to break a bond, the electron moves to the conduction band. There is now an unoccupied level in the valence band. This unoccupied state, or incomplete covalent bond, has a positive charge associated with it, is easily dissociated from the original site, and becomes free to take part in conduction, much the same as the electron. This positively charged quasi particle is called a *hole*. So when a bond is broken, an electron–hole pair (EHP) is produced. Such EHPs can also be produced by absorption of a photon of sufficient energy, which is the basis for solar cells and a variety of photodetectors.

---

[1] These hybrid states are denoted by sp^3 because one of the s electrons is promoted to a p level and four distinct orbitals are formed.

[2] An eV (electron-volt) is the kinetic energy gained by one electronic charge when accelerated through 1 volt. 1 eV $= 1.6 \times 10^{-19}$ J.

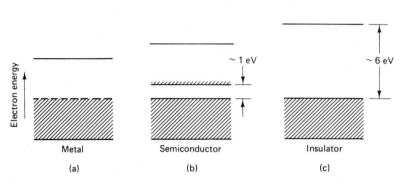

Figure 2-2 Simplified energy band scheme for metals, semiconductors, and insulators.

Up to this point the nature of pure semiconductors has been discussed. Such semiconductors are called *intrinsic* semiconductors because the properties are intrinsic to the material. Most interesting properties of semiconductors arise from the addition of impurities. The latter semiconductors are referred to as *extrinsic*. To illustrate the addition of impurities to semiconductors, silicon (Si) will be used as an example. Silicon forms the diamond structure by sharing its four valence electrons with four surrounding atoms in a tetrahedral coordination. Some Si atoms in the crystal can be substituted for by atoms of similar size, such as boron (B) or phosphorus (P), which have valences of 3 and 5, respectively. When P is substituted, the one electron left over is easily dissociated from the parent atom to become a free electron in the conduction band. The P atom becomes a negative fixed ion, and no hole is produced in the process. Since the number of intrinsic EHPs at room temperature is about 1.5×10 cm^{-3} and the number of atoms about 5×10^{22} cm^{-3}, even a light impurity concentration of one part per million (ppm), or 5×10^{16} cm^{-3}, means that there will be about 1 million more extrinsic electrons than intrinsic EHPs. Thus the predominant charge carrier will be electrons, and this extrinsic material is labeled n type for negative-charge carriers. Actually, the number of holes diminishes markedly because of the law of mass action, which requires that

$$np = n_i^2 \tag{2-1}$$

where n is the number of electrons per unit volume, p the number of holes per unit volume, and n_i the number of electrons or holes per unit volume in intrinsic material. Thus, for the example given,

$$p = \frac{(1.5 \times 10^{10})^2}{0.5 \times 10^{16}} = 4.5 \times 10^4$$

or 12 orders of magnitude less than n. The holes can be ignored.

If Si is doped with B, then a *p*-type (positive-charge carriers) material is produced by the same reasoning as for the *n*-type material. The B atom has a valence of 3 and cannot complete the four bonds with its neighbors. The incomplete bond is easily dissociated from the parent B atom, making it a negative ion and producing a hole in the valence band capable of electrical and thermal conduction.

The properties of *p*- and *n*-type semiconductors by themselves are of limited value. However, when combined in pairs, that is, *pn* junctions, numerous electronic and optical devices becomes possible.

2-3 *pn* JUNCTIONS

Metallurgical junctions between *p*- and *n*-type materials are produced by diffusion, epitaxial growth, and ion implantation. In all cases, the junction is made in a single crystal. It is instructive, however, to imagine starting with two separate crystals, one *n* type, the other *p* type, and suddenly joining them to form one continuous single crystal with the metallurgical junction at the interface. Figure 2-3 depicts the various steps in this thought experiment. At the instant the materials are joined there is an infinite concentration gradient across the junction for both holes and electrons. Consequently, holes and electrons *diffuse* in opposite directions, producing a large current until sufficient fixed ionic charges are uncovered in the newly created depletion region (devoid of free charge) to produce a counterelectric field that causes an opposing drift current. The following current balance holds true at equilibrium.

$$I_{\text{dif elec}} = I_{\text{drift elec}} \tag{2-2}$$

$$I_{\text{dif holes}} = I_{\text{drift holes}}$$

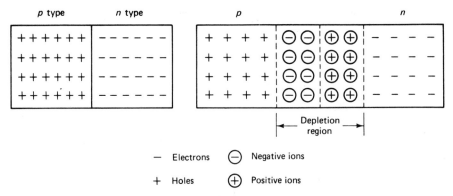

Figure 2-3   *pn* junction through experiment. When the junction is formed, holes and electrons diffuse across the junction, uncovering bound ionic charge in the depletion region. The resulting built-in electric field creates drift currents that balance the diffusion currents.

Equations (2-2) state that at equilibrium the diffusion current of electrons equals the drift current of electrons and that the same is true for holes. In terms of the energy band scheme, this situation can be represented as in Fig. 2-4, where only the lower conduction band and upper valence band edges are shown. E_F is called the Fermi energy, and for our purposes it is a reference energy that lies about in the middle of the energy gap for intrinsic materials and is shifted toward the valence band for p-type and toward the conduction band for n-type materials. E_F is constant throughout a semiconductor at equilibrium; thus it aids in drawing energy-level diagrams.

The electric field built into the depletion region as a result of the uncovered ions produces a potential energy barrier for electrons and holes, preventing substantial diffusion of either charge carriers from their majority areas into the minority regions. The pn-junction device being considered here is a diode and can be either forward or reverse biased by placing voltages across it, as depicted in Fig. 2-5. When the diode is forward biased,

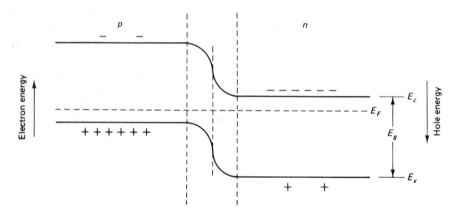

Figure 2-4   Energy band edges for pn junction at equilibrium.

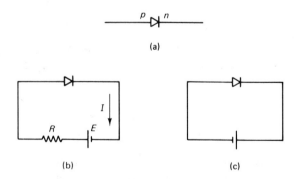

Figure 2-5   (a) Diode symbol. (b) Forward-biased diode. (c) Reverse-biased diode.

the Fermi level is shifted upward on the *p* side and downward on the *n* side, thus lowering the potential energy barrier and allowing a large increase in the *diffusion* of holes and electrons across the junction. A resistor is placed in this circuit to limit current. The drift current is decreased, but this factor has relatively little effect because it is a current of minority carriers and there are few to begin with. When the diode is reverse biased, the shifts in the Fermi level are reversed, thereby increasing the potential energy barrier. Although there is a small increase in the drift current, it is relatively unimportant because it is a minority current and cannot be increased to any very large absolute value. In reverse bias the diode acts like an extremely high resistance. The energy-level picture of forward and reverse biasing is given in Fig. 2-6. The diode current versus diode voltage curve (characteristic curve) is sketched in Fig. 2-7 for Si. Note that the current scales are different for forward and reverse voltages.

A Si diode is said to *turn on* at about 0.55 to 0.7-V forward bias; 0.6 V will be used here for convenience. Large increases in current occur for

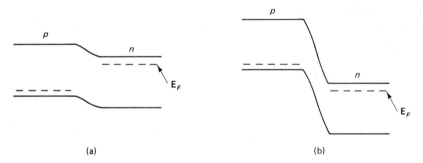

Figure 2-6   Energy band edges for (a) forward-biased and (b) reverse-biased diodes.

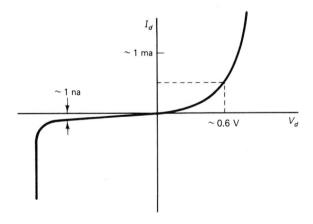

Figure 2-7   Diode characteristic curve for Si, showing reverse breakdown.

voltages slightly larger than the turn-on voltage. The reverse current is relatively constant and usually less than 1 nA until reverse breakdown occurs. This phenomenon will be discussed subsequently. The characteristic curve, exclusive of the breakdown region, obeys the following diode equation closely:

$$I = I_0 \left(\exp \frac{eV}{kT} - 1 \right) \tag{2-3}$$

where I_0 is the reverse saturation current, e the electronic charge, V the voltage across the diode, k Boltzmann's constant, and T absolute temperature. A convenient number to remember is that $kT = 0.026$ eV at $T = 300$ K.

The reverse breakdown is caused by two phenomena. At low voltages it is caused in extremely heavily doped diodes by quantum mechanical tunneling of electrons through a very thin depletion layer. At higher voltages in less heavily doped diodes, the cause is an avalanche effect in which electrons passing through the depletion layer acquire sufficient energy, because of the large electric field produced by the reverse bias, to ionize neutral atoms in the depletion layer. At a sufficiently high voltage the result can be an avalanche or multiplying effect. Neither type of breakdown is inherently harmful. Damage occurs if the heat generated by a large breakdown current is not dissipated rapidly enough to prevent melting of the diode. Reverse breakdown is used for voltage control and as the basis of avalanche photodiodes.

Different types of semiconductor diodes have different turn-on voltages and reverse saturation currents. Some values of these quantities are listed in Table 2-1.

In determining the resistor size to use in a circuit in order to limit the forward current through a diode, it is customary to assume a voltage drop across the diode equal to its turn-on voltage. This situation is illustrated in Fig. 2-8 for a Si diode. The resistance in Fig. 2-8 is given by

$$R = \frac{12 \text{ V} - 0.6 \text{ V}}{0.050 \text{ A}} = 288 \ \Omega$$

TABLE 2-1 Reverse Saturation Currents and Turn-on Voltages for Some Important Semiconductor Diodes

	Reverse Saturation Current	Turn-on Voltage (V)
Si	<1 nA	0.6
Ge	<1 μA	0.3
GaAs	—	1.2
GaAs$_{0.6}$P$_{0.4}$	—	1.8
GaP	—	2.2

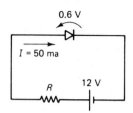

Figure 2-8 Silicon diode circuit for current limitation.

To limit the current during reverse breakdown, we would do the same as in Fig. 2-8 except that the precise value of the reverse breakdown voltage must be known. This value can be obtained accurately by means of a curve tracer.

2-4 TRANSISTORS AND SIMPLE CIRCUITS

Many different types of transistors are available. In this section the bipolar junction transistor (BJT) is discussed, and the junction field-effect transistor (JFET) and metal oxide semiconductor (MOS) are briefly described. The BJT is referred to as such because two types of charge carriers are responsible for its properties. It has two back-to-back *pn* junctions, as schematically depicted in Fig. 2-9, for an *npn* transistor.

The regions of the BJT are designated E for emitter, B for base, and C for collector. The base region is very thin, approximately 1 μm. The depletion regions are shown shaded. The doping levels become progressively lighter going from emitter to collector, with typical values of 10^{19} cm^{-3}, 10^{17} cm^{-3}, and 10^{15} cm^{-3} for emitter, base, and collector, respectively. The transistor symbol and an instructive, although impractical, circuit are shown in Fig. 2-10.

Referring to the circuit in Fig. 2-10, the emitter–base junction is forward biased, thus causing a large flow of electrons into the base region with

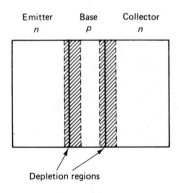

Figure 2-9 Schematic of *npn* transistor.

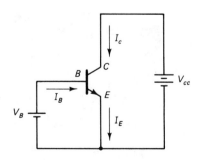

Figure 2-10 Simple circuit illustrating BJT biasing.

a relatively small flow of holes into the emitter because of the lighter doping in the base. The base region is so thin that most electrons pass directly through it into the electric field of the reverse-biased base–collector junction, which carries them on into the collector. A small reverse saturation current occurs at the base–collector junction. The few electrons that do not make it through the base, plus the hole current from the base to emitter, are made up for by a small flow of electrons out of the base. The currents shown in Fig. 2-10 are conventional currents; so they are in the opposite direction of the electron flow. According to Kirchhoff's current law, the three currents must be related as follows:

$$I_E = I_B + I_C \tag{2-4}$$

The dc or small-signal current gain of the transistor is $\beta = I_C/I_B$, so

$$I_E = \left(1 + \frac{1}{\beta}\right) I_C \tag{2-5}$$

Because β usually has a value of 50 to 400, it is frequently sufficient to assume $I_E = I_C$. pnp transistors behave similarly except that the current directions and voltage polarities are reversed. Whenever possible, transistors are made *npn* to take advantage of better conductivity of electrons compared to holes.

Most silicon BJTs are constructed in a thin epitaxial layer on a silicon chip, and so the actual device looks more like the sketch in Fig. 2-11, although it is still highly schematic. Contact to *E, B,* and *C* is made by means of aluminum deposited through openings in a SiO₂ passivation layer, which is grown directly on the Si. The n^+ symbol represents doping approaching the solid solubility limit. The buried layer is used to reduce the resistance (collector resistance) between the collector and collector contact. The draw-

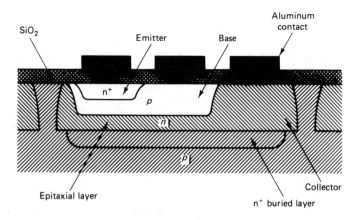

Figure 2-11   Silicon BJT.

ing, of course, is not to scale. The lateral dimensions are reduced by at least a factor of 10 compared to the vertical direction. Obviously some transistors, particularly high-power devices, may look much different than Fig. 2-11, but this diagram is fairly representative of BJTs used in integrated circuits, low-power applications, and optoelectronic devices.

Transistor circuits are covered thoroughly in many excellent books. It may be helpful to the uninitiated reader, however, to present some simple circuits here.

The BJT may be used either as an amplifier or as a switch. A simple switching scheme is shown in Fig. 2-12. The transistor does not turn on until the voltage drop across the $E–B$ junction approaches 0.6 V. In Fig. 2-12, $V_{in} = 0.6$ V $+ I_E R$ and $V_{out} = I_E R$. Another useful configuration in logic circuits is the inverter shown in Fig. 2-13. When $V_{in} = 5$ V, V_{out} is near 0 V, and when $V_{in} = 0$, $V_{out} = 5$ V. R_1 can be used to limit the base current, and R_2 drains stored charge out of the base when the transistor is turned off to improve the speed of operation. The bleed-off resistor technique has been supplanted in integrated circuits through the use of Schottky diode clamped transistors. The Schottky diode is a metal-silicon diode that turns on at about 0.3 V and is built from collector to base in such a way that the $B–C$ junction cannot become forward biased, which is what causes charge to be stored in the base when the transistor is turned on.

Figure 2-14 contains two simple biasing schemes for small-signal, one-stage amplification. These circuits are frequently useful with optoelectronic devices. In Fig. 2-14(a), the voltage gain is $v_{out}/v_{in} = -R_c/r_e$, where $r_e = 0.026$ V$/I_E$, and in Fig. 2-14(b) the voltage gain is $v_{out}/v_{in} = -R_c/R_E$. The minus sign is used because of the phase change seen in the inverter, and v_{out} and v_{in} are small-signal ac voltages. The biasing resistors R_1 and R_2 are chosen to make $I_1 \approx I_2 \geq 10I_B$, so

$$V_B = V_{cc} \frac{R_2}{R_1 + R_2} = 0.6 \text{ V} + I_E R_E$$

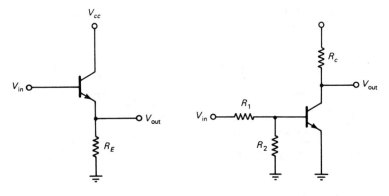

Figure 2-12   BJT used as a switch.   **Figure 2-13**   BJT used as an inverter.

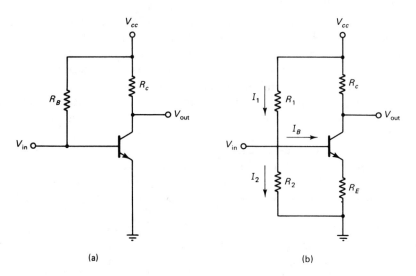

(a)   (b)

Figure 2-14   Simple common-emitter amplifiers.

where I_1, I_2, and I_E are dc currents. If the output is taken from the emitter, the amplifier is referred to as an emitter follower. The voltage gain is slightly less than 1, but the output resistance is extremely low, making the emitter follower very desirable as the output stage of a multistage amplifier, such as an operational amplifier (op amp).

The purpose of the biasing resistors R_B in Fig. 2-14(a) and R_1, R_2 in Fig. 2-14(b) is to provide a dc quiescent or operating point for the transistor. Then the small ac signals input to the base will be amplified at the output. One of the biasing resistors may be replaced by a photodiode or phototransistor in optical detection schemes.

A JFET is schematically depicted in Fig. 2-15. When a voltage is applied between drain D and source S as shown, conduction takes place by means of a drift current caused by the electric field produced in the channel

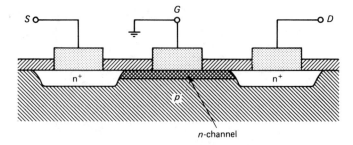

Figure 2-15 n-channel JFET.

between D and S. If V_D becomes large enough, or a negative voltage is applied to the gate G, the channel begins to pinch off as a result of an increase in the size of the depletion layer between the p and n regions. The current saturates at this point. BJTs are essentially linear in that, when properly biased, changes in output current are linearly proportional to changes in input current. JFETs obey a quadratic relationship; in this case, $I_D \propto V_G^2$, where V_G is voltage applied to the gate. This nonlinearity makes FETs less desirable for linear amplifiers, but it has many useful applications in communications and optical detection. JFETs have high input resistances, for the gate current is always looking into a reverse-biased diode, relative to the drain. That is, $I_s = I_G + I_D$ and I_G is extremely small. This makes JFETs useful in the first stage of high-input resistance amplifiers, such as some of the better op amps and the first amplifier stage in radios.

A MOSFET device is shown in Fig. 2-16. In these transistors, conduction is by field effect, that is, drift current, just as in JFETs. The gate, however, is completely insulated from the semiconductor. These devices have also been dubbed insulated gate FETs or IGFETs. In the version shown in Fig. 2-16, a negative voltage applied to the gate pushes electrons out of the channel, thus depleting it, and this normally on device is turned off at high enough negative gate voltage. Like JFETs, the current voltage relation is quadratic. The input resistance of these devices is extremely high, because the gate is insulated. MOS devices are used heavily in digital circuits and in some optical devices.

Because of its usefulness in many optical applications, a brief introduction to the ideal op amp is given here. The reader is referred to any of a large number of electronics books for more detailed coverage. Some simple optical applications of op amps will be presented later.

The ideal op amp has infinite input resistance, zero output resistance, and infinite open-loop gain. To a first approximation, any amplifier can be modeled by the circuit shown in Fig. 2-17. The ideal amplifier would then have $R_{in} = \infty$, $R_0 = 0$, and $A_0 = \infty$. This means that v_{in} can come from a high-resistance source without losing voltage to the source resistance and

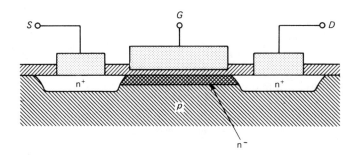

Figure 2-16   MOS n-channel depletion-mode device.

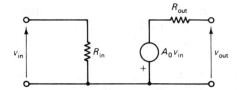

Figure 2-17   Circuit model for amplifier.

the output of the amplifier can be connected to a low resistance load without losing voltage in R_0.

Having $v_0 = Av_i = \infty$ is a convenience that will become apparent as we look at some simple op-amp circuits. The symbol used to represent an op amp and two useful circuits, an inverting amplifier and a noninverting amplifier, are presented in Fig. 2-18. Because of the large open-loop gain

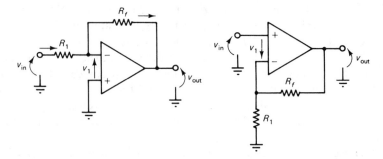

Figure 2-18   (a) Inverting amplifier. (b) Noninverting amplifier.

(this means no feedback loop) of the op amp, v_1 in Fig. 2-18(a) is nearly zero, thus making the negative or inverting input a virtual ground. Also, because of the high input resistance, the current between the + and − terminals is essentially zero. Consequently,

$$i_1 = \frac{v_{in} - 0}{R_1} = \frac{0 - v_{out}}{R_f}$$

or

$$A_c = \frac{v_{out}}{v_{in}} = -\frac{R_f}{R_1} \qquad (2\text{-}6)$$

which is the closed-loop gain. The input resistance of this amplifier is R_1 and its output resistance is[3]

$$\frac{R_0}{A_0} \frac{R_1 + R_f}{R_1}$$

---

[3] Do not confuse op-amp input and output resistance with amplifier input and output resistance.

In Fig. 2-18(b) the roles of the inverting and noninverting inputs are reversed. Now the voltage at the inverting terminal is approximately v_{in}. Thus

$$i = \frac{v_{in} - 0}{R_{in}} = \frac{v_{out}}{R_{in} + R_f} \tag{2-7}$$

and so

$$A_c = \frac{v_{out}}{v_{in}} = \frac{R_1 + R_f}{R_1}$$

The gain is a little higher and the input and output voltages are in phase. The input resistance of this amplifier is $A_0 R_{in} (R_1/R_1 + R_f)$, which is generally very large, and its output resistance is the same as the inverting amplifier.

The input voltage in these circuits may come from a photodetector or all or part of the input resistance may be replaced by a photoresistor.

A standard low-cost op amp, the 741, has $R_{in} = 1$ to 2 MΩ, $R_0 = 75$ Ω, and a dc open-loop voltage gain of -2×10^5 that decreases with frequency to -100 at 10 kHz. For the relations just given to be valid, the open-loop gain of the op amp must always be much greater than the closed-loop gain.

PROBLEMS

2-1. The energy of the ground-state electron in hydrogen is -13.6 eV. In the first excited state the energy is -3.4 eV. Calculate the frequency and wavelength of the light given off when an electron undergoes a transition from the first excited state to the ground state.

2-2. Write out the electronic configurations in spectroscopic notation for Mg, Si, P, and K.

2-3. A crystal of n-type silicon at room temperature has 5×10^{18} phosphorus atoms per cubic centimeter. How many electrons and holes are there per cubic centimeter?

2-4. The turn-on voltage for a GaAs light-emitting diode (LED) is 1.2 V. What resistance is needed in series with the LED to provide a current of 50 mA if a 12-V battery is used?

2-5. A transistor has a β of 200. Calculate the emitter current if the collector current is 1.5 mA. What is the base current?

2-6. Design a simple inverting op-amp circuit to have a closed-loop gain of 15.

2-7. Design a noninverting op-amp circuit to have a gain of 10.

2-8. The energy gap in GaAs is 1.4 eV. Calculate the wavelength and frequency of the photon given off when an electron from the conduction band recombines with a hole from the valence band.

REFERENCES

HEWLETT-PACKARD, *Optoelectronic Designer's Catalog*. Palo Alto, CA: 1978.

KITTEL, C., *Solid State Physics*, 5th ed. New York: John Wiley & Sons, Inc., 1976.

TILL, W. C., AND J. T. LUXON, *Integrated Circuits: Materials, Devices, Fabrication*. Englewood Cliffs, NJ: Prentice Hall, 1982.

VANVLACK, L. H., *Elements of Materials Science and Engineering*, 4th ed. Reading, MA: Addison-Wesley Publishing Co., 1980.

Chapter 3

Radiometry, Photometry, Optical Device Parameters, and Specifications

This chapter introduces radiometric and photometric quantities and the parameters and physical characteristics of solid-state light emitters and detectors that are used to describe the physical properties most frequently considered when deciding which device is best suited for a given application. The spectral output, power and/or brightness, spatial distribution of light, current requirements and limitations, and voltage requirements are discussed for light emitters. Responsivity R, detectivity D, detectivity per unit frequency D^*, noise equivalent power NEP, spectral response, and frequency response are covered for photodetectors. Some comments about noise considerations in photodetectors are made, but an in-depth coverage of this subject is beyond the scope of this book.

3-1 RADIOMETRY AND PHOTOMETRY

When the response of the eye is considered in the measurement or specification of visible radiation, the subject is called *photometry* as opposed to *radiometry*, in which direct physical measurements or specifications are made. The eye response curve is given in Fig. 3-1 for what is called photopic or day vision. The curve is shifted toward shorter wavelengths for night or

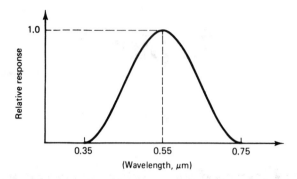

Figure 3-1 Spectral response curve of the eye (color response).

scotopic vision. The peak of the curve occurs at 555 nm for photopic vision and at 510 nm for scotopic vision. Photopic receptors in the eye are sensitive to color and light level, whereas the scotopic receptors are sensitive only to light level. The scotopic receptors provide what is therefore referred to as *night vision*. Not everyone, of course, has the same eye spectral response, but Fig. 3-1 is accepted as representative. The unit of luminous flux is the lumen, corresponding to the radiometric unit, the watt. Incident power per unit area (W/cm²) in radiometric terms is called *irradiance*. In photometric terms it is referred to as *illuminance* (lumens/cm²). The units given here are not necessarily SI units, but are commonly used combinations. The proper SI units adhere strictly to the MKS system where applicable, such as in area units.

To calculate luminous flux for 1 W of radiant flux at a given wavelength, you must multiply by $685V(\lambda)$, where $V(\lambda)$ is the luminous efficiency obtained from the eye spectral response curve (CIE luminous efficiency curve). The product, $685V(\lambda)$, is known as the *luminous efficacy*. When the radiation consists of a spread of wavelengths, spectral quantities are used. These are actually radiant or luminous quantities per unit wavelength. The spectral radiant flux has the units of watts per meter. To obtain the total radiant flux, we must integrate the spectral radiant flux over wavelength. To obtain the total luminous flux, we would integrate $685V(\lambda)$ times the spectral radiant flux over wavelength, as shown in Eq. (3-1).

$$\Phi_v = 685 \int_\lambda V(\lambda)\Phi_e(\lambda) \, d\lambda \qquad (3\text{-}1)$$

In this equation the subscript v denotes a visual or luminous quantity, whereas the subscript e denotes an electromagnetic or radiant quantity.

Before discussing the remaining photometric and radiometric quantities, the concept of solid angle must be introduced. The solid angle Ω is the three-dimensional counterpart of the usual planar or two-dimensional angle defined by the ratio of the arc of a circle to the radius of the circle, as illustrated in Fig. 3-2. A solid angle is defined as the ratio of the area of a portion of a sphere to the radius squared. This concept is illustrated in Fig. 3-3. The full planar angle is $2\pi r/r = 2\pi$ radians, whereas the full solid angle is $4\pi r^2/r^2 = 4\pi$ steradians. A useful relation between θ and Ω exists for small angles and circular beams. Refer to Fig. 3-4 for an explanation of this relationship. As can be seen, $\Omega \approx \pi\theta^2/4$ in this case.

Luminous intensity and radiant intensity are defined as luminous flux per unit solid angle and power per unit solid angle, respectively. A lumen per steradian is called a *candela*. These quantities define the ratio of the flux emanating from a source into a cone, defined by solid angle Ω to the solid angle of that cone. A laser emits a beam of very high intensity compared with most thermal light sources because the beam it emits has an extremely small divergence angle. The radiant intensity of a 1-mW HeNe laser is 1000 W/sr for a full-angle beam divergence of 1 mrad.

The final quantity of interest is intensity per unit-emitting area. In photometry it is called *luminance* and might have units of candelas/cm^2. The radiometric term is *radiance*, and the units are frequently W/cm^2-sr. Luminance (radiance) is the flux emitted by an area of the source divided by the emitting area and the solid angle of the cone into which the flux is emitted. Luminance is often referred to as brightness.

Consider a 1-mW HeNe laser with a full-angle divergence of 1 mrad and a beam diameter of 1 mm. The radiance for this laser will be approximately 1.6×10^5 W/cm^2-sr. It is informative to compare it with the radiance of the sun, which is about 130 W/cm^2-sr, and a 1000-W mercury arc lamp, which has a radiance of 1000 W/cm^2-sr [Ready, 1971]. The radiance (luminance) of a source cannot be increased by any optical means; it can only be decreased by stops or attenuation.

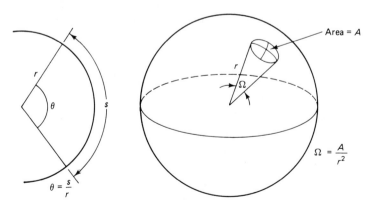

Figure 3-2 Definition of planar angle.

Figure 3-3 Definition of solid angle.

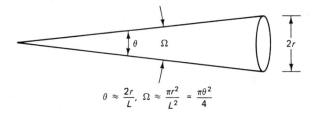

$$\theta \approx \frac{2r}{L}, \ \Omega \approx \frac{\pi r^2}{L^2} = \frac{\pi \theta^2}{4}$$

Figure 3-4 Relation between θ^2 and Ω for small-divergence angle, circular cross-sectional beam.

3-2 PROPERTIES OF LIGHT EMITTERS

Certain *pn*-junction diodes emit light when forward biased. They are referred to as light-emitting diodes (LEDs). Like any other diode, a certain forward turn-on voltage must be provided, and the current through the device can be limited by a series resistor. LED manufacturers specify the turn-on voltage and maximum allowable current. Light output versus current is nearly linear for LEDs, and this curve may be supplied with the other specifications. LEDs cannot tolerate a very large reverse voltage because of low-voltage breakdown due to heavy doping. The tolerable reverse voltage is given in the specifications.

LEDs emit relatively monochromatic, although incoherent, radiation with wavelength bandwidths of 100 Å. Typical wavelengths are 0.9, 0.66, and 0.55 μm in the ir, red, and green, respectively. The power output in watts for infrared LEDs is usually given. The spatial distribution of the power output may be important and is given in terms of a plot of irradiance versus angle, such as in Fig. 3-5.

The output of visible LEDs may be given in terms of intensity or brightness. Many LEDs, particularly those used in displays, approximate lambertian sources. A lambertian source or diffuser emits radiation with an

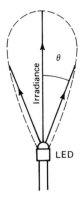

Figure 3-5 Irradiance versus angle for an LED.

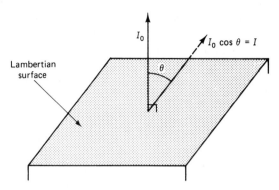

Figure 3-6   Variation of I with angle for a Lambertian surface.

intensity variation given by

$$I = I_0 \cos \theta \qquad (3\text{-}2)$$

where I_0 is the intensity normal to the surface and θ is the angle between the radiating direction and the normal. Figure 3-6 depicts this situation. A reflecting or transmitting surface of this sort is considered an ideal diffuser of light. Brightness (radiance) is constant for any viewing angle because the projected area of the surface as viewed from any direction decreases as $\cos \theta$. This factor is important for displays intended for wide-angle viewing. It is left as an exercise to show that $I_0 = \pi^{-1} \times$ (total flux from the surface) for a lambertian surface. So, for an approximate lambertian emitter, any two of the three (I_0, flux F, or brightness B) can be calculated if the third is given.

3-3 QUALITY FACTORS FOR PHOTODETECTORS

Solid-state photodetectors that convert incident radiation into a measurable electrical current or voltage are compared by the use of several defined parameters. The simplest is the responsivity R, which is the ratio of the output signal (in amperes or volts) to the incident irradiance. The responsivity of a photodiode, for example, might be 1 μA/mW/cm^2 compared to 1 mA/mW/cm^2 for a phototransistor. The spectral responsivity R_λ is the responsivity per unit wavelength and is generally plotted as a function of wavelength to provide the spectral response of the device, or the peak value is given along with the wavelength at which it is maximum. For photodiodes, the quantum efficiency η is sometimes given. It is simply the number of EHPs generated per incident photon. Responsivity is proportional to quantum efficiency.

The detectivity of a device, D, is the ratio of the responsivity to the noise current (voltage) produced in the detector. Thus it is the signal-to-noise (SN) ratio divided by irradiance. The units, therefore, are simply cm²/ W. This parameter is little used, but serves to introduce the following parameter, D^*, which is the detectivity times square root noise frequency bandwidth, $(\Delta f)^{1/2}$, times square root detector area, $A^{1/2}$. The reason for defining D^* as

$$D^* = D(A\ \Delta f)^{1/2} \tag{3-3}$$

is to eliminate the dependence on noise bandwidth and detector area. Noise power is proportional to the frequency bandwidth and detector area, and so the noise current or voltage will be proportional to $(A\ \Delta f)^{1/2}$. The units of D^* are cm³ Hz$^{1/2}$/W or cm Hz$^{1/2}$/W if power instead of irradiance is used in the definition of R. D^*_λ, the spectral form of D^*, may be given in addition to D^* or in place of it.

Noise equivalent power (NEP) is the reciprocal of D and is therefore simply the irradiance required to produce a signal-to-noise ratio of 1. The most widely accepted figure of merit is D^*.

The values of any parameters mentioned depend on how they are measured. Measurement conditions are usually specified. For example, D^*(500 K, 400, 1) means that D^* was measured with a 500 K black-body source chopped at 400 Hz, using 1-Hz bandwidth. Actually, the bandwidth is larger but is normalized to 1 Hz.

Noise considerations. The main sources of noise in semiconductor photodetectors are thermal noise and generation–recombination noise. Random fluctuations in the thermal motion of electrons will cause an rms current, given by

$$i_n = \sqrt{\frac{4kT\ \Delta f}{R}} \tag{3-4}$$

The random fluctuations in generation and recombinations of EHPs cause an rms current, given by

$$i_n = \sqrt{4eIG\ \Delta f} \tag{3-5}$$

where G is the photoconductive gain, which will be discussed later.

A variety of noise sources is loosely referred to as one over f noise because all are proportional to $\sqrt{\Delta f}/f$ and hence are only important at low frequency. Because noise powers add directly, the total noise current is given by the square root of the sums of the squares of the individual noise currents. Note that in all cases the noise current is proportional to $(\Delta f)^{1/2}$.

Sources of radiation are also sources of noise. At low light levels, statistical variations in the rate of arrival of photons at the detector cause what is called *quantum noise*. Fluctuations in the interference of waves are a source of thermal noise in a detector. Another source of noise arises from randomly varying capacitance due to vibrations or breakdown of dielectric films.

3-4 SPECTRAL RESPONSE

The spectral response is generally presented as a variation of responsivity as a function of wavelength. Spectral response characteristics for specific types of detectors are described in Chapter 4, but some general comments are appropriate here. Photodetectors fall into two general categories: thermal detectors and quantum detectors. Examples of thermal detectors are calorimeters, thermopiles, bolometers, thermistors, and pyroelectric detectors. In all these devices the incident radiation is converted to heat, which, in turn, causes a temperature rise and a corresponding measurable change, such as an emf generated in a thermocouple or a current pulse in a pyroelectric detector. Detectors of this type tend to have a flat spectral response over a wide wavelength range.

Examples of quantum detectors are photomultiplier tubes, semiconductor photodiodes, and phototransistors. In this type of detector, photons produce a measurable change directly, such as the emission of electrons in photomultiplier tubes or the generation of EHPs in semiconductor photodiodes. Photoemissive surfaces for photomultipliers have been designed with fairly flat spectral response but have a sharp cutoff at the long wavelength end, where the photon energy is insufficient to overcome the work function of the material. Semiconductor detectors have fairly narrow spectral response, peaked at the wavelength corresponding to photons with energy equal to the band gap energy.

3-5 FREQUENCY RESPONSE

The frequency response of a photodetector refers to its ability to respond to an incident light beam chopped or modulated at various frequencies. Most solid-state detectors behave like low-pass filters and as such their responsivity can be written as

$$R = \frac{R_0}{(1 + \omega^2 \tau^2)^{1/2}} \tag{3-6}$$

where R_0 is the responsivity at zero frequency and τ is the time constant. The cutoff frequency is $f_c = (2\pi\tau)^{-1}$, where $R = R_0/\sqrt{2}$. The time constant may be given by $\tau = rC$, where C is the device capacitance and r is device resistance plus circuit resistance. Clearly, an inherently fast device (wide frequency response) can be slowed down by use of a large circuit resistance.

PROBLEMS

3-1. Calculate the solid angle for a HeNe laser that emits a beam with a full-angle divergence of 1 mrad if the diameter of the emitted beam is 2 mm.

3-2. The total flux (power) from a lambertian surface is 2 mW. Calculate the intensity at normal incidence and the brightness (radiance) if the surface is 1 mm square.

3-3. Calculate the noise current in a 100-Ω resistor at room temperature for a bandwidth of 1 kHz.

3-4. The dc responsivity of a solid-state photodetector is 10 mA/mW with a time constant of 10^{-7} s. What is the responsivity at 10^8 Hz?

3-5. Name and briefly discuss the reasons for three types of noise common to photodetectors.

3-6. Name the figures of merit used to qualify photodetectors. Write out the equations for them and explain.

REFERENCES

CHAPPEL, A. (ed.), *Optoelectronics: Theory and Practice*. The Netherlands: Holland Printing Partners. Texas Instruments Ltd., 1976.

DRISCOLL, W. G. (ed.), AND W. VAUGHN (assoc. ed.), *Handbook of Optics*. New York: McGraw-Hill Book Co., 1978.

PEDROTTI, F. L., AND L. S. PEDROTTI, *Introduction to Optics*. Englewood Cliffs, NJ: Prentice-Hall, 1987.

RCA, *Electro-Optics Handbook*. Harrison, NJ: RCA Commercial Engineering, 1968.

READY, J. F., *Effects of High-Power Radiation*. New York: Academic Press, Inc., 1971.

WILLIAMS, W. S., AND O. A. BECKLUND, *Optics: A Short Course for Engineers and Scientists*. New York: John Wiley & Sons, Inc., 1972.

Chapter 4

Light-Detection Devices

This chapter describes a variety of light-detection devices that are frequently used in laser applications. The discussion includes energy and power measurement as well as a brief description of imaging devices, such as photodiode arrays and charge-coupled devices (CCDs). Some simple detector circuits are also briefly described.

4-1 CLASSIFICATION OF DETECTORS

Light detectors can be placed in either of two classifications: thermal or quantum (sometimes referred to as photon). Thermal detectors are those in which the absorbed light energy is converted into heat that produces a temperature rise that, in turn, causes some detectable output or variation in a property of the detector. Quantum detectors are those in which the absorbed light directly alters a property or produces a measurable output.

Examples of thermal detectors are thermocouples, calorimeters, bolometers, and pyroelectric devices. The radiation causes a resistance change in bolometers; either a voltage or a current pulse is produced in the others. In all cases, it is a change in temperature that produces the measurable phenomenon.

Quantum detectors are exemplified by photoemissive devices, such as photomultiplier tubes (PMTs), photodiodes, photoconductors, and phototransistors. A resistance change is directly produced in photoconductors by the absorption of light; in the others a measurable current or voltage is generated by direct interaction of the absorbed light with electrons.

Generally, thermal detectors are slower and have a broader spectral response than quantum detectors. There are notable exceptions to this statement, however, particularly with respect to the speed of response.

4-2 THERMAL DETECTORS

As noted, thermal detectors are based on a change in property or output that results from a temperature increase arising from the absorption of light. In most devices a certain degree of thermal inertia is inherent because heat conduction is involved. Consequently, the response time of such detectors tends to be very slow unless the mass is kept small, as in some thin-film devices. Most thermal detectors use a blackened absorbing surface and frequently exhibit relatively flat spectral response from the uv to the far ir. Window materials placed in front of the absorbing surface and the absorbing coating determine spectral response.

Calorimeters.   Although many different types of calorimeters exist, only a few are briefly described here to elucidate the principles involved.

Several calorimeter designs use an absorbing cavity that is suitably coated for high absorption and may have a conical shape to produce multiple reflections of the light entering the cone. The temperature changes is sensed by thermocouples or thermistors[1] embedded in the absorber, which is usually aluminum or copper because of their large thermal conductivity. Such a device is schematically represented in Fig. 4-1.

When light is absorbed by such a device for a period of time, the absorbed energy causes a uniform temperature increase in the highly conductive Al or Cu. This temperature increase is sensed by the thermocouples or thermistors and may be displayed as a voltage on a meter or digital readout. Calibration is carried out by irradiation with a standard source or by supplying a known quantity of heat through a resistive heater embedded in or attached to the absorber. Such calorimeters can measure power by being allowed to reach thermal equilibrium with their surroundings or coolant. The detected steady-state temperature rise is then a measure of the incident power. Calibration would occur in the same way as for energy measurement.

Because the response time of these devices may be several seconds, it is possible to measure high power levels by chopping the beam and allowing

---

[1] A thermistor is a ceramic resistor with a large temperature coefficient of resistance.

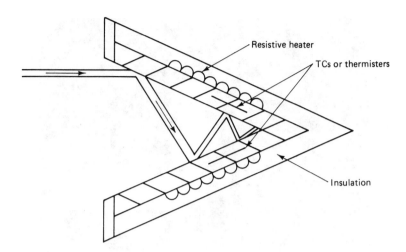

Figure 4-1 Cone-type calorimeter.

only a small fixed percentage (for example, 1% to 3%) of the power to enter the calorimeter. In addition, because of the inherent thermal inertia of such devices, the output signal will remain relatively steady. Power levels of thousands of watts can be measured in this fashion. Figure 4-2 is a photograph of a commercial calorimeter that uses an absorption cavity for high-power CO_2 laser power and/or energy measurement.

Some calorimeters are water cooled, and the difference in the inlet and outlet water temperature is used as a measure of the power or energy absorbed. Because heat capacity C is defined as heat absorbed per unit mass per unit temperature change, Eq. (4-1) gives the power P absorbed in terms of heat capacity, temperature change ΔT, and mass flow rate dm/dt:

$$P = \frac{dQ}{dt} = C \, \Delta T \, \frac{dm}{dt} \qquad (4\text{-}1)$$

A "rat's nest" calorimeter uses wire, such as copper, randomly tangled inside a highly reflective container. The tangled mass (or mess if you prefer) of wires ensures nearly 100% absorption of the incident radiation, and the temperature rise is sensed as an increase in the resistance of the wire. Absolute calibration is achieved by dissipating a known amount of electrical power in the wire. Because a thermally induced resistance change is used, this device could also be called a rat's nest bolometer.

When thermistors or bolometers are used to sense temperature changes, it is possible to place two identical calorimeters in a bridge arrangement for very sensitive power or energy measurements. Ambient conditions are the same for both calorimeters, except that one is irradiated and the other is not. This technique is depicted in Fig. 4-3. The variable resistance is adjusted for a null reading on the galvanometer. When the calorimeter is

Figure 4-2   Absorption cavity calorimeter. (Courtesy of Scientech, Inc.)

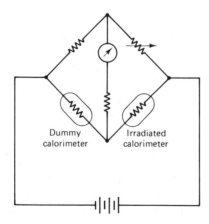

Figure 4-3   Bridge network for calorimeter.

illuminated, the bridge is unbalanced, thus causing a deflection of the galvanometer that can be calibrated in terms of power or energy for maximum deflection. The identical dummy calorimeter tends to provide cancellation of resistance change effects due to variations in environmental conditions, which should nominally be the same for both calorimeters.

Thermopiles.   A thermopile consists of a number of thermocouples connected in series. A thermocouple is a junction of two different metals or semiconductors. Because of the work function difference between the materials, a current is generated when the junction is heated. Thermocouples use a reference junction, and the voltage difference between the junctions is a measure of the temperature difference between them. The reference voltage in modern systems is established electronically so that only one junction, or probe, is required. Such devices are extremely common in temperature measurement. When a junction of dissimilar materials, with an appropriate absorbing coating, is irradiated, a temperature increase occurs, thus generating a voltage that is a measure of the power absorbed.

The voltage produced in metallic thermocouples is only a few microvolts per degree of change in temperature, and so it is common practice to arrange several couples in series so that all the junctions of one polarity are irradiated. In this way, their voltages add to give a larger response, the reference junctions being masked from the radiation. Figure 4-4 is a sketch of such a thermopile. The thermopile may be incorporated as one leg of a bridge circuit with an identical dummy device in the other leg to improve sensitivity and immunity to environmental fluctuations. Thin-film thermopiles can have response times on the order of microseconds with a flat spectral response.

Bolometers.   Bolometers were discussed earlier and so little will be added here. Such devices are either metallic or semiconductor resistors in

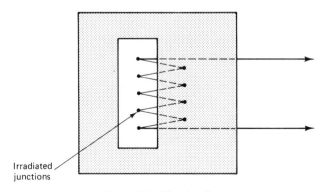

Irradiated junctions

Figure 4-4   Thermopile.

which a thermal change in resistance is induced by absorbed radiation. The change in resistance causes a change in current in a circuit and hence a voltage change across the bolometer or another resistor in the circuit. Bolometers, of course, can be used in a bridge circuit with a dummy bolometer in one leg.

Pyroelectric detectors.   Pyroelectric detectors are made from materials that have permanent electric dipole moments associated with them, such as $LiNbO_3$, triglycine sulfate (TGS), and polyvinyl fluoride (PVF). These materials are ferroelectric, meaning that permanent dipole moments are associated with the unit cell, in the case of crystalline materials, or the basic molecular unit in the case of polymers. These dipoles are all aligned parallel within domains analogous to the domains in ferromagnetic materials. The chief difference between pyroelectric ferroelectrics and other ferroelectrics is the difficulty of poling pyroelectrics. To get the saturated polarization of all domains aligned in the same direction, the material must be placed in a strong electric field at an elevated temperature, and the polarization cannot be reversed at normal temperatures by application of an electric field. The total dipole moment of pyroelectric materials is, however, very sensitive to temperature fluctuations.

When absorbed by pyroelectric detector, light is converted to heat, thereby causing a temperature rise and hence a change in polarization. Figure 4-5 is a simplified pyroelectric detector scheme. Charges absorbed on the surface neutralize the polarization surface charge. When the polarization is altered due to a temperature change, charge must be transferred through the external circuit, for it cannot be desorbed from the surface quickly. The speed of response for such devices is very good, in the nanosecond range, because the speed is not thermal conductivity (thermal inertia) limited. The temperature change causes a change in dipole moment that responds extremely fast.

In the circuit of Fig. 4-5, R should be less than the resistance of the detector, which is very high because it is basically a capacitor. The voltage

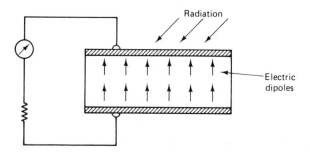

Figure 4-5   Simplified pyroelectric detector scheme.

developed by the detector is given by

$$V(t) = AR\gamma \frac{dT}{dt} \qquad (4\text{-}2)$$

where γ is the pyroelectric coefficient or polarization change per unit temperature change and A is the irradiated area.

Pyroelectric detectors have a flat spectral response to long wavelengths without the need for cryogenic cooling required by other ir detectors, and their frequency response, which obeys Eq. (3-5), is very good, with bandwidths around a gigahertz. Maximum incident power is limited to a few tens of watts per square centimeter. Pyroelectric arrays are used for imaging laser beams for a wide range of wavelengths.

4-3 QUANTUM DETECTORS

Devices that produce a response as a direct result of the absorption of radiation are referred to as *quantum* or *photon detectors* and, indeed, some photoemissive devices can detect single photons.

Photoconductors.   When a photon is absorbed by a semiconductor, the probability that an electron–hole pair will be produced is very high. The quantum efficiency η, which is defined as the ratio of conduction electrons produced to absorbed photons, is usually 0.7 or higher at the peak of the spectral response curve. A photoconductor is an intrinsic or extrinsic semiconductor in which the change in conductivity resulting from the absorption of light is used as a measure of the incident power. Unlike a bolometer, temperature change has nothing to do with the conductivity change. Conductivity is given by

$$\sigma = ne\mu_e + pe\mu_h \qquad (4\text{-}3)$$

where n and p are the number of electrons and holes per unit volume, respectively, and μ_e and μ_h are the mobilities of electrons and holes, respectively. The creation of electron–hole pairs in an intrinsic conductor alters both n and p. The long-wavelength cutoff of the device is given by

$$\lambda_c = \frac{hc}{E_g} \qquad (4\text{-}4)$$

where E_g is the width of the energy gap, h is Planck's constant, and c is the speed of light. The signal current, assuming the dark resistance if very large, is given by

$$i_s = e\eta A \frac{\Delta f t_c}{t_t} \qquad (4\text{-}5)$$

where A is the irradiated area, Δf is the bandwidth, t_c is the carrier lifetime, and t_t is the transit time (time for an electron to traverse the device). The ratio t_c/t_t is called the *photoconductive gain* because i_s can be increased by designing the photoconductor to have long carrier lifetime and short transit time. The main source of noise in photoconductors is generation–recombination noise, which also increases with t_c/t_t; so the signal-to-noise ratio is not improved by increasing the photoconductive gain.

Narrow bandgap photoconductors, such as HgCdTe, are used extensively for ir detection but must be cryogenically cooled to minimize the effect of thermally generated charge carriers.

Photodiodes.   Semiconductor diodes were discussed in Chapter 3, and so the basic theory of diodes need not be considered here. When a semiconductor diode is used as a photodetector, an internal current is produced as a result of the absorption of photons and the creation of electron–hole pairs (EHPs). Many types of photodiodes and modes of operation are possible. The more common types are discussed next.

All *pn*-junction devices respond to incident radiation if the junction is close enough to the surface for light to reach it and if the wavelength of the incident radiation is less than the cutoff wavelength. Longer wavelengths tend to penetrate farther, and short wavelengths are absorbed nearer the surface. The uv response of photodiodes is poor both because EHPs generated near the surface tend to encounter surface recombination centers and because most of the energy of the uv photons is wasted, for only a small fraction is needed to produce an EHP.

The process by which current is generated can be understood by reference to Fig. 4-6. It can clearly be seen that an EHP generated in the depletion region will tend to produce a current as the election and hole slide down their respective potential hills. If the diode has a resistor connected across it, a current is produced as long as light is incident on the junction. This situation is depicted in Fig. 4-7. This is the photovoltaic mode of op-

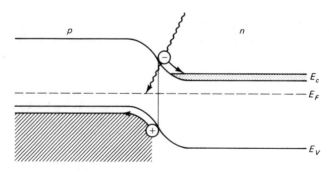

Figure 4-6 Energy-level diagram for an unbiased photodiode.

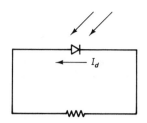

Figure 4-7 Photovoltaic mode of operation.

eration, for a voltage is produced across the resistor and power $V_d I_d$ is produced.

Actually, EHPs need not be generated exactly in the depletion region. EHPs generated within one diffusion length on either side of the junction will contribute to the current. The diffusion length is the average distance that an electron or hole travels in the minority regions before recombining. The diffusion length for holes is substantially less than that for electrons; so most diffused diodes are made with a thin *p*-type region on top (toward the incident radiation).

The ideal responsivity of photodiodes is easily deduced in the following way. The photocurrent is given by

$$I_p = \eta e \frac{dn}{dt}$$

where dn/dt, the rate at which photons strike the detector, is given by P_i/bf. Here P_i is the incident power. The responsivity in amperes per watt is then

$$R = \frac{I_p}{P_i} = \frac{e\lambda}{hc} = 8.1 \times 10^5 \lambda \quad (\text{A/W}) \tag{4-6}$$

which yields a typical responsivity of about 0.6 A/W at $\lambda = 1.0 \ \mu m$.

Photodiodes may have areas of 1 cm^2 or greater for high sensitivity or position detection, but this factor increases noise and capacitance. Many photodiodes are very small, having an active area of less than 0.01 cm^2. The responsivity for small-area devices is more conveniently defined as output current per unit irradiance in W/cm^2. A typical responsivity is 1 mA/W/cm^2.

The major source of noise in photodiodes is shot noise, which is caused by the tendency of electrons and holes to bunch up or to cross the depletion region in a random manner.

The short wavelength response of photodiodes can be enhanced by placing the junction closer to the surface. Overall response is then sacrificed, of course. To achieve improved response at all wavelengths in the spectral range of the device, *pin* diodes are used. Here an intrinsic layer (actually a very lightly doped region) is placed between the *p* and *n* regions. This step improves quantum efficiency by providing a larger depletion region and increases speed of response by decreasing junction capacitance.

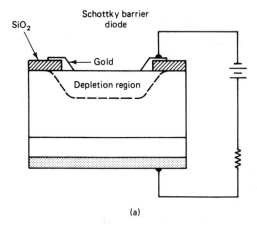

(a)

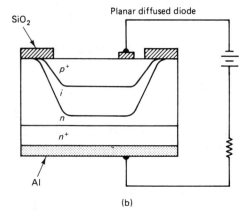

(b)   **Figure 4-8** Silicon photodiode designs.

Two common silicon photodiode designs are the planar-diffused and Schottky barrier diodes. Figure 4-8 contains sketches of both types. The planar-diffused type may be made as a *pin* or ordinary *pn* diode, depending on the spectral response and speed requirements. The Schottky barrier diode has a superior uv response, due to the fact that the depletion region extends to the silicon surface. Large-area Schottky diodes are less noisy than their planar-diffused counterparts but have lower power-handling capability.

The diode equation can be generalized to account for photogenerated current. If a short circuit is placed across an illuminated diode, the resulting diode current is a reverse current and is called the *short-circuit current*, I_{sc}. If a resistor is placed across the illuminated diode, then a forward-biased voltage is produced and the diode current is given by

$$I_d = I_0(e^{eV/kT} - 1) - I_{sc} \qquad (4\text{-}7)$$

Note that the open-circuit voltage, V_{oc}, is given by

$$V_{oc} = \frac{kT}{e} \left(\ln \frac{I_{sc}}{I_0} + 1 \right) \tag{4-8}$$

Figure 4-9 is a sketch of diode current versus diode voltage for various incident light levels. When a diode is operated in the photovoltaic mode, it is operating in the fourth quadrant of the current versus voltage graph. The short-circuit current varies nearly linearly with incident power over about six decades. So a circuit that provides essentially zero voltage drop across the diode provides substantially linear operation.

Maximum power output in the photovoltaic mode, $(V_d I_d)_{max}$, is given by a point determined by inscribing a rectangle of maximum area as shown in Fig. 4-10. The process involves matching the load resistance to the diode resistance. Unfortunately, the diode resistance varies with incident light level in the photovoltaic mode. Consequently, no single resistance will work for all light levels, and an optimum load based on operating conditions must

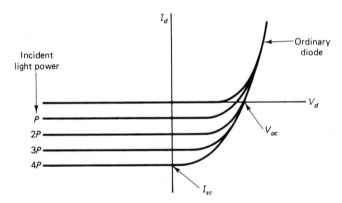

Figure 4-9   Diode current versus diode voltage for various light levels.

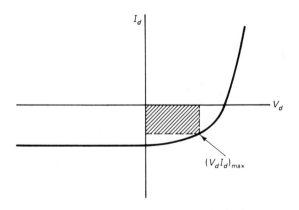

Figure 4-10   Maximum power output point.

be determined. Solar cells and nonbattery-powered light meters operate in this fashion.

When a photodiode is reverse biased, the operation may be linear over 10 or 11 orders of magnitude change in light level and the junction capacitance is reduced, thereby decreasing the response time in some diodes to less than 1.0 ns. Reverse-biased operation is referred to as the *photoconductive mode*, but it should not be confused with the term photoconductor; photoconductors are not even diodes.

Avalanche photodiodes are reverse biased at a precisely controlled voltage that is slightly less than the breakdown voltage. When light strikes the diode, creating EHPs, the avalanche process begins and continues as long as the light is incident on the diode. The advantage of this type of diode is the multiplication factor. It may be as great as 1000, which provides an internal gain and correspondingly high responsivity. The response time of avalance diodes is also very good, about 1.0 ns.

Phototransistors.   A phototransistor is just what the name implies— a transistor used as a light detector to take advantage of the internal gain of the transistor. Light incident on the base–collector junction of a bipolar transistor produces an internal current that acts as the base current of an ordinary transistor. Consequently, no base lead is required unless external biasing is necessary.

The spectral response of silicon phototransistors is similar to that of silicon diodes, but their responsivities are about three orders of magnitude larger, typically around 1 $A/W/cm^2$. The response time of phototransistors is microseconds and slower.

Many phototransistors are built in a Darlington configuration to take advantage of the increased current gain realized from such a design. The light is allowed to impinge on the collector–base junction of the first transistor, and the second one provides additional gain. The responsivity of these devices is typically 10 $A/W/cm^2$ or higher.

Photo-FET devices are available for applications in which the very high input resistance of such transistors is required. Photo-SCRs are also available for light-activated switching applications, such as fire alarms.

Photon drag.   Detectors using the principle of photon momentum transfer, referred to as *photon drag detectors*, are not extensively used in manufacturing settings, but they are potentially useful for on-line monitoring of CO_2 laser power and measurement of the time dependence of CO_2 laser pulses.

The momentum of a photon can be determined by equating the mass–energy equivalence relation to the photon energy,

$$mc^2 = hf$$

so that momentum, $p = mc$, is given by

$$p = \frac{hf}{c} = \frac{h}{\lambda} \tag{4-9}$$

In such materials as germanium and tellurium, which are transparent to 10.6-μm radiation, a sufficient amount of momentum is transferred from the photons to electrons or holes to cause a measurable open-circuit voltage across a crystal through which the radiation is passing. This open-circuit voltage is the result of the holes or electrons being pushed or dragged along with the radiation. Response time may be less than 1.0 ns, with responsivities of 10^{-8} to 10^{-6} V/W.

Photomultipliers.   If the energy of photons incident on an absorbing surface exceeds the work function[2] of the electrons in that surface, the photoelectric effect is observed. The incident photons eject electrons for the surface by direct transfer of energy. Collection of these electrons at an anode produces a photoelectric current that is a measure of the light power incident on the photoemissive cathode.

Photodetectors that use a single photoemissive surface and a single collector are simply called *phototubes* or *photodiode tubes*. Phototubes are very fast but have low responsivities because there is no internal gain mechanism. In photomultipliers a series of *dynodes* is used to provide multiplication or gain. The electrons ejected from the cathode are accelerated by a voltage of 100 to 200 V to a dynode where they, on impact, eject many secondary electrons, which are, in turn, accelerated to another dynode, and so on. Tubes may contain ten or more dynodes and produce a multiplication factor of 1 million or more. Response times (transit time of the electrons) of less than 1.0 ns are achieved and single photons can be detected, thus leading to the term photon counting, even though not every photon incident on the emissive surface is detected.

A variety of photocathodes covers the spectral range from 0.1 to 11.0 μm. Examples are (1) Ga–As, which has a fairly flat spectral response from 0.13 to 0.8 μm with a peak responsivity of around 35 mA/W, and (2) Ag–O–Cs alloy, which covers the spectral range from 0.3 to 1.0 μm, although the responsivity varies significantly.

Many photomultiplier designs exist. Figure 4-11 is a schematic representation of one type of design. Photomultiplier tubes are extremely fast and sensitive but expensive, and they require sophisticated associated electronics to operate.

---

[2] Work function is the energy required to remove an electron from the surface of the material to infinity.

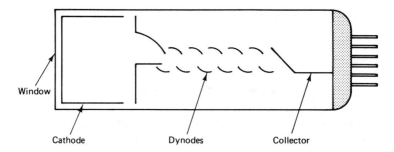

Figure 4-11   Photomultiplier tube.

Position detectors.   A variety of silicon photodiode position detection devices is available. Such devices fall into two categories: multidiode units and lateral-effect diodes.

Multidiode units consist of either two or four separate, but identical, photodiodes placed on the same silicon wafer but electronically isolated from each other. The bicell and quadrant detectors are schematically depicted in Fig. 4-12. When a symmetrical laser beam is centered on either detector, the output of each diode is identical. If the beam moves more onto one diode than the other(s), the signal from that diode increases while the signal from the other(s) decreases. This change in output can be used to measure the displacement of the laser beam. Measurements of beam displacement of as little as 1.0 μm are claimed to be possible. One-directional displacements can be measured with the bicell, whereas two-dimensional displacements can be measured with quadrant detectors. The linear displacement of the beam on the detector can also be easily related to angular displacement of the beam.

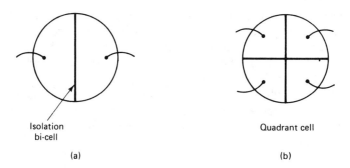

Figure 4-12   Photodiode position dectors.

The beam profile generally used with position detectors is a filtered Gaussian distribution. Consequently, the output of the diodes is not linearly related to the amount of beam displacement. Detectors of this type can be made either by the planar-diffused process as either *pn* or *pin* diodes or by the Schottky barrier process.

The lateral-effect diode position detector avoids the problem of nonlinear output as a function of beam displacement. Figure 4-13 is a schematic diagram of a lateral-effect diode. The output current from terminal *S* is ideally given by

$$I_s = I_0\left(1 - \frac{S}{L}\right) \qquad (4\text{-}10)$$

where I_0 is the actual photogenerated current across the depletion layer at the point of incidence of the beam. The farther the beam is from the output terminal, the more electrons cross back over the depletion layer and recombine at the surface. This type of position detector is slower and noisier than multicell types, but the change in output with beam position is very linear, and the beam irradiance profile is irrelevant. Beam displacements as small as 0.01 μm can be detected with silicon diode position detectors, and this factor allows very small angular changes to be detected.

Figure 4-14 shows a variety of silicon photodetectors that can be used for power measurement, centering, and position measurements.

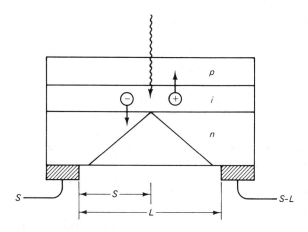

Figure 4-13 Lateral-effect diode. (Adapted with permission of Laser Focus.)

Figure 4-14   Silicon photodetectors.

4-4 DETECTOR ARRAYS

A thorough discussion of detector arrays is well beyond the scope of this book. A brief description of the basic types of arrays is included, however, because of their usefulness in lower-power laser applications. Linear and area arrays using photodiodes, charge-injection devices (CIDs), and charge-coupled devices (CCDs) are available. Detector arrays are frequently used in systems designed to measure part dimensions and surface location.

Photodiode arrays.   As the name implies, photodiode arrays consist of either one- or two-dimensional arrays of photodiodes. The output of each diode is proportional to the amount of light incident on it; so imaging is possible with a resolution dependent on the cell (pixel) size and center spacing of the pixels. In this context, a pixel refers to the basic light-detection unit.

The light-sensitive area is about half the total pixel area. These devices are usually back illuminated so that there are no contact obstructions. Figure 4-15 is a schematic illustration of a linear photodiode array. Arrays of this type may have from 64 to 2048 elements on center-to-center spacings as low as 2.5 μm. Light incident on the silicon photodiodes produces charge that is stored in the parallel capacitor until a voltage from the shift register is applied to the gate of the FET. The charge is then dumped onto the video line so that a sequence of pulses whose amplitudes are proportional to the amount of light energy incident on the diode during the framing period is read out. The slower the clock rate is the more charge can be stored and the more sensitive the device is to light level. Sampling rates may be as high

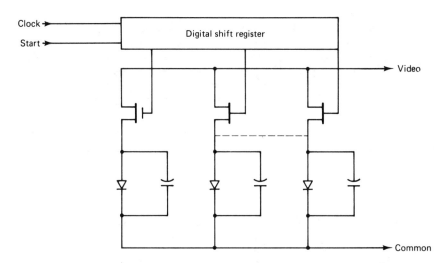

Figure 4-15   Linear photodiode array. (Adapted courtesy of EG & G Reticon Corp.)

as 40 MHz. These devices also come with CCD-analog shift registers. In this version a voltage is applied simultaneously to the gates of all the FETs, the charge stored in each capacitor is dumped into the CCD shift register, and the charge packets are then read out serially as with the digital shift register. In both types of arrays the output is a discrete time analog of the light irradiance distribution across the array.

Two-dimensional diode arrays with up to 256 × 256 elements are available. In these devices a shift register sequentially applies pulses to the FET gates row by row, dumping the charge from each diode in a row into one or two CCD shift registers, which then serially output the charge packets onto one or two video lines. Pixel rates of 10 MHz can be achieved.

Charge-coupled devices.   CCDs use metal oxide semiconductor (MOS) capacitors directly to store packets of charge produced by the incident light. The charge packets, which represent an integration of the light incident on a particular capacitor, are then serially shifted out of the CCD by transference from one capacitor to another. Simplified versions of two schemes for achieving CCD imaging are shown in Fig. 4-16.

In three-phase CCD imagers, the charge is collected under the gates of capacitors that have a negative gate voltage applied, for instance, θ_1. As a result, a depletion region is created under the θ_1 gate oxide into which holes generated by incident photons are attracted and stored. The number of holes in the packet is proportional to the number of photons incident during the period of time that θ_1 is negative. The charge packets can then be transferred to the θ_2 capacitors by reducing θ_1 and making θ_2 large, and

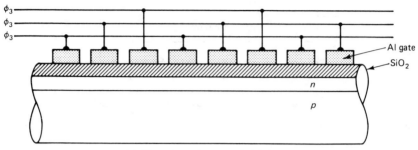

(a) Three-phase charge-coupled device

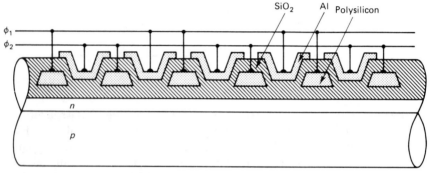

(b) Two-phase charge-coupled device

Figure 4-16 CCD linear imagers. (Adapted with permission from *Solid State Technology*, Technical Publishing, a company of Dun and Bradstreet.)

then to capacitor θ_3 by making θ_1 and θ_2 small and θ_3 large, and so on, until the charge is read out as an analog pulse similarly to the output of a photodiode array. Alternatively, the packets stored in the imaging capacitors can be transferred initially to a parallel line of MOS capacitors that is masked so that light cannot reach them, and then the charge is transferred to the output in the manner described.

To prevent the trapping of electrons by fast surface states at the oxide interface, a *fat zero* (a small number of holes) can be cycled through the capacitors continually to keep these states filled. This process minimizes the degradation of the signal. Another technique is to use ion implantation to put a shallow p-type buried layer below the oxide; doing so pushes the depletion layer down into the bulk Si away from the surface states and keeps the signal charge away from the surface state traps.

Charge transfer in the right direction in three-phase devices is determined by the asymmetry in the application of the clock voltages, θ_1, θ_2, and θ_3. Charge transfer in the right direction can also be accomplished in a two-phase system by introducing a built-in asymmetry. Figure 4-16(b) illustrates one way of doing so. The oxide thickness under the poly-Si gates is less than under the Al gates; thus the oxide capacitance for the poly-Si gate capacitors is higher than that of the Al gate capacitors. Consequently, a smaller voltage occurs in the depletion region beneath the Al gate than below the poly-Si gate capacitors. This situation can be considered in terms of potential wells: the wells under the Al gates are always less deep than the wells under the corresponding poly-Si gates. If θ_1 is large, charge is accumulated in the θ_1 capacitors. As θ_1 is decreased and θ_2 increased, charge can only move to the right because of the potential barrier that exists on the left due to the shallower well under the Al gate.

Transfer efficiencies, the amount of charge transferred from one MOS capacitor to the next, of over 99.999% are possible. Center-to-center spacing of pixels is typically 12.7 μm and a dynamic range[3] of 1000 to 1 is common. The elements in CCD imagers can be made about as small as the limits of photo or electron beam lithography will allow. Very large linear and area arrays that can operate at both high- and low-light-level conditions are possible. It has been reported that 200 million pixels per second can be read in large displays through the use of multiple readout taps.

Charge injection devices.   CID imagers are similar to CCDs in that photon-generated charge is collected and stored in MOS capacitors. In two-dimensional imagers, each pixel consists of two side-by-side capacitors. When a gate voltage is applied to one, it gathers and stores charge. The voltage is then transferred to the second capacitor, which is masked from the light, by application of the voltage to it and removal of the voltage from the first. Readout is accomplished by sequentially setting the voltages to zero for a line that has been selected by placing all the second (storage) capacitor voltages high. When the voltage goes to zero, the charge is injected into the substrate, thus causing a detectable current. Using this technique, it is possible to address a two-dimensional array on an individual basis (random access) by applying appropriate column and row voltages. A linear array does not require the second set of capacitors unless individual addressing is necessary.

4-5 PHOTODETECTOR CIRCUITS

Circuits for the operation and amplification of photomultiplier tubes, photodiode arrays, and CCD or CID arrays are beyond the scope of this book. Such circuitry can be found in the applications notes provided by the manufacturer. A few simple but useful circuits for photodiodes, phototransistors, or photoconductors are described, however.

---

[3] Ratio of highest to lowest detectable light levels.

It is instructive to see how a photodiode behaves under reverse-biased conditions. Care must be taken to choose the correct load resistance and bias voltage if the diode response is to be linear. The circuit is depicted in Fig. 4-17.

If the normal characteristic curves for a photodiode are inverted and rotated 180° about the y axis, they look similar to transistor characteristic curves, as shown in Fig. 4-18, where I_p is the photocurrent and V is the voltage across the diode. If I_p is sufficiently large, the diode becomes forward biased and the output is nonlinear in response to light input power. Applying Kirchhoff's voltage law to the circuit in Fig. 4-17 yields

$$I_p - \frac{1}{R} V_{cc} - \frac{1}{R} V \qquad (4\text{-}11)$$

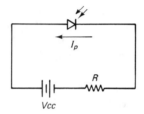

Figure 4-17   Reverse-biased photodiode.

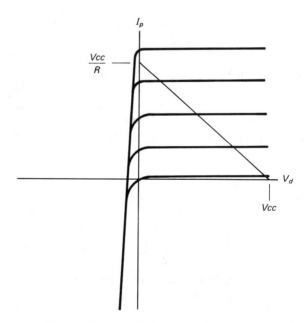

Figure 4-18   Photodiode characteristic curves.

It is readily seen that if the absolute value of the slope is too small or V_{cc} is too small, V may become negative (this actually means forward biased in this convention). This is also referred to as saturation since further increases in light level produce negligible changes in V. Proper choice of R and V_{cc} will produce a quiescent (operating) point such that variations in light level about some nominal value result in linear variations in V.

A simple light-activated relay circuit is depicted in Fig. 4-19. Phototransistors can handle sufficient current to activate small relays. When light strikes the phototransistor, turning it on, the relay is opened or closed, depending on its normal state. When the light is removed, the relay returns to its normal state. The diode may be placed in parallel with the relay to protect it from possible voltage spikes.

A simple phototransistor circuit of the sort shown in Fig. 4-20 can be used for switching or detection operations and even for light-modulation applications in which true fidelity of the modulation wave form is not critical. Voice communication on an amplitude-modulated laser beam, for example, can be accomplished by taking the output of this phototransistor amplifier and feeding it into an audio amplifier.

A transistor circuit that can be used to amplify a photodiode output is given in Fig. 4-21. The photodiode in Fig. 4-21 may be replaced by a photoconductor.

An operational amplifier circuit for photodiodes that provides great linearity and high speed is depicted in Fig. 4-22. In this circuit, E_c and R_2 may be zero. E_c is provided to reverse bias the diode, which produces faster and more linear operation. The resistance R_1 may be chosen to balance offset current resulting from diode dark current. The output voltage for this circuit is given by

$$v_0 = i_p R_1 \qquad (4\text{-}12)$$

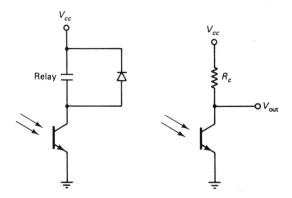

Figure 4-19 Light-activated relay.   **Figure 4-20** Phototransistor amplifier.

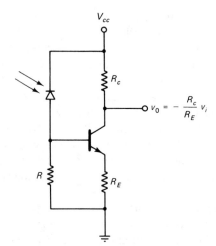

Figure 4-21 Transistor amplifier circuit for a photodiode.

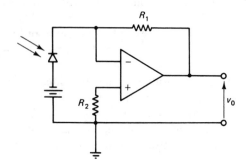

Figure 4-22 Op-amp circuit for linear photodiode operation.

In applications where large light-level changes may be encountered, a logarithmic amplifier, such as the one given in Fig. 4-23, may be useful. In this case, the op amp must have an extremely high input impedance. It can be obtained by using an FET input stage. Here the op amp is used in a noninverting amplifier arrangement, where

$$v_0 = \left(1 + \frac{R_1}{R_2} \right) v_i$$

Because of the very high input impedance, $I_d \simeq 0$ so that

$$0 = I_0(e^{eV_d/kT} - 1) - I_{sc}$$

which leads to

$$V_d = v_i = \frac{kT}{e} \ln \left(1 + \frac{I_{sc}}{I_0} \right)$$

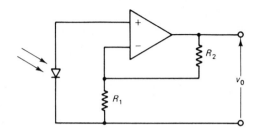

Figure 4-23 Logarithmic amplifier for photodiode applications.

Thus the output voltage is given by

$$v_0 = \left(1 + \frac{R_1}{R_2}\right)\frac{kT}{e}\ln\left(1 + \frac{I_{sc}}{I_0}\right) \tag{4-13}$$

If energy measurement is desirable, a simple op-amp integrating circuit (shown in Fig. 4-24) may serve the purpose. In this circuit the output voltage is proportional to the integral of the input voltage, which is, in turn, proportional to the power incident on the photodiode. Therefore, the output voltage is a measure of the energy absorbed by the diode during the time interval of the integration. Equation (4-14) is the relation between output voltage and input voltage.

$$v_0 = \frac{1}{RC}\int_0^t v_i \, dt \tag{4-14}$$

4-6 BEAM PROFILERS

Beam quality is important in most laser applications whether they employ a 1-mW or a 10-kW laser. Patterns burned into photosensitive paper, cardboard, wood, or plastic are useful for alignment and determining beam symmetry, but they do not give complete information concerning the irradiance profile.

For lasers that operate in the ir or visible region of the spectrum, CID, CCD, or diode arrays can be used to determine the beam profile, assuming

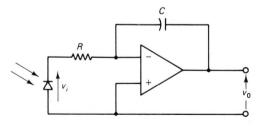

Figure 4-24 Integrating circuit for energy measurement.

the beam is adequately attenuated to prevent damage or saturation of the devices. For CO_2 lasers, pyroelectric arrays are utilized for their long-wavelength response. With suitable processing by a computer, a three-dimensional plot of the irradiance profile can be obtained.

Obtaining beam profile information at or near focus is impractical for standard detector arrays. A method that has achieved some success in this application is a scanning wire profiler. In this method a wire is rapidly scanned through the beam, and light scattered from the wire is detected. By scanning wires in orthogonal directions, the beam profile can be generated.

PROBLEMS

4-1. (a) Name the two classes of photodetectors, describe them, and explain how they work.

(b) List three types of detector in each class and explain why they are classified that way.

4-2. What should the mass flow rate in a water-cooled calorimeter be if a 1°C temperature change is to occur for each 100 W of power absorbed?

4-3. Assuming the specific heat of a pyroelectric detector is C and its mass is m, determine an expression relating output voltage, area, series resistance, pyroelectric coefficient, the given quantities, and the power P incident on the device.

$$\textit{Answer:} \quad V = AR\gamma \frac{P}{mc}$$

This assumes all the power is absorbed by the detector and none is lost by conduction, convection, or radiation.

4-4. (a) Calculate the conductivity of an n-type silicon with $N = 2 \times 10^{18}$ cm^{-3}:

$$\mu_e = 1300 \text{ cm}^2/\text{V} \cdot \text{s}$$

(b) Calculate the long wavelength cutoff for silicon.

$$E_g = 1.1 \text{ eV}$$

4-5. Calculate the ideal responsivity in amperes per watt for a photodiode operating at $\lambda = 0.66$ μm.

4-6. Calculate the open-circuit voltage at room temperature for a photodiode if $I_0 = \ln A$ and $I_{sc} = 100$ mA.

4-7. Name three different types of semiconductor photodetectors and explain their operation.

4-8. Describe two distinct types of diode position detectors and compare them.

4-9. Name two types of detector arrays and briefly explain how they operate.

4-10. Design a sample op-amp circuit that will provide linear output and an output of -1 V/μA of photodiode current.

4-11. In a simple reverse-biased photodiode detection scheme, what resistance is required to provide a quiescent point in the midrange if the bias voltage is 5 V and the photocurrent is 50 μA?

REFERENCES

AMELIO, G. F., "Charge-Coupled Devices," *Scientific American*, February 1974, 21–31.

DRISCOLL, W. G. (ed.), AND W. VAUGHN (assoc. ed.), *Handbook of Optics*. New York: McGraw-Hill Book Co., 1978.

Engineering Staff of Texas Instruments, Inc., Semiconductor Group, *The Optoelectronics Data Book for Design Engineers*, 2nd ed. Dallas, TX: Texas Instruments, Inc., 1975.

General Electric Co., Optoelectronic Systems Operation, *CID, Theory of Operation and Random Access Description*. Syracuse, NY: General Electric Co.

MELEN, R., AND D. BUSS (eds.), *Charge Coupled Devices: Technology and Applications*. New York: IEEE Press, 1977.

MOSS, T. S., G. J. BURRELL, AND B. ELLIS, *Semiconductor Opto-Electronics*. New York: John Wiley & Sons, Inc. 1973.

STIMSON, A., *Photometry and Radiometry for Engineers*. New York: John Wiley & Sons, Inc., 1974.

STOTLAR, S. C., AND E. J. McLELLAN, "Developments in High-Speed Pyroelectric Detectors," *Optical Engineering*, 21, No. 3, May–June 1981, 469–71.

STREETMAN, B. G., *Solid State Electronic Devices*, 2nd ed. Englewood Cliffs, NJ: Prentice Hall, 1981.

Chapter 5

Introduction
to the Laser

The purpose of this chapter is to present an overall view of lasers and their applications. In this way, it is hoped that the reader will have a clearer and more cohesive understanding of the material presented in subsequent chapters. It may also enable to reader to be selective, if desired, in the choice of reading from the remaining chapters. The reader who has some familiarity with lasers and the various types available may wish to skip this chapter.

The acronym LASER is derived from the expression Light Amplification by Stimulated Emission of Radiation. Used in this sense, the term refers to electromagnetic radiation anywhere in the spectrum from the ultraviolet (uv) through the infrared (ir), which includes wavelengths from roughly 0.1 to 1000 μm. The term MASER was originally coined to describe a similar device (using ammonia molecular transitions and a microwave cavity) that operates in the microwave region of the electromagnetic spectrum, in this case, at a wavelength of 1.25 cm.

The acronym, laser, although slightly redundant, is highly descriptive. Light and radiation, in this context, are the same thing. The key words are "amplification" and "stimulated emission." It is the ability of light to stimulate the emission of light that creates the situation in which light can be amplified. It may be helpful to point out at this time that a very important feature of most lasers is an optical resonator, usually consisting of two precisely aligned mirrors, one of which is partially transmitting, to allow an output. This mirror arrangement provides positive feedback. So the laser is basically a positive feedback oscillator.[1] As such, it is analogous to electrical positive feedback oscillators where a certain amount of the output is fed back in phase with the input, resulting in oscillation at some frequency characteristic of the circuit. In effect, the oscillator selects a frequency component from the noise always present from biasing, amplifies it, and oscillates at that frequency. The laser does essentially the same thing except that an optical oscillator can operate in many allowed modes (natural resonator frequencies).

5-1 UNIQUE PROPERTIES OF LASER LIGHT

The laser is basically a light source. The radiation that it emits is not fundamentally different than any other form of electromagnetic radiation. The nature of the device, however, is such that some remarkable properties of light are realized. These unique properties, taken as a whole, are not available from any other light source to the extent that they are obtained from a laser. The unique properties referred to are the following.

1. High monochromaticity (small wavelength spread)
2. High degree of both spatial and temporal coherence (strong correlation in phase)
3. High brightness (primarily due to small beam divergence)
4. Capability of very low (microwatts) to very high (kilowatts) continuous power output for different types of lasers
5. High peak power (terrawatts) and large energy (hundreds of joules) per pulse in pulsed output lasers
6. Capability of being focused to a small diffraction-limited spot size (on the order of the wavelength of the light)

These properties are by no means independent of each other and, in fact, some may be inferred directly from others.

5-2 REQUIREMENTS FOR LASER ACTION

A number of conditions must be satisfied to achieve lasing action. They are listed here and are briefly discussed at this point, with more detail following later.

---

[1] In some cases, such as the N_2 laser, the gain is sufficiently great that feedback is not required to achieve lasing action.

1. Population inversion
2. Optical resonator, except in extremely high gain systems
3. Lasing medium
4. Means of excitation
5. Host medium (usually)

The notion of a population inversion refers to a condition in which a certain ensemble of atoms or molecules is in a nonequilibrium situation, where more of these atoms or molecules are in some specified excited energy state (electronic or vibrational) than are in a lower-energy state. These atoms or molecules undergo a transition to a lower state in which the probability of emission of a photon is extremely high, a radiative transition.

The optical resonator refers to the technique for providing positive feedback into the system to produce oscillation. This process usually consists of two parallel mirrors placed some distance apart; one is as nearly totally reflecting as possible and the other is partially transmitting to obtain a useful output from the system. The transmittance of the output mirror, which is the ratio of transmitted power to incident power, ranges roughly from 1% to 60%, depending on the power level and type of laser.

The lasing medium (called the *lasant*) refers to the atoms or molecules that actually emit the light, such as Ne atoms in a HeNe laser or Cr^{3+} ions in a ruby laser.

Some means of excitation is required to achieve the population inversion. This step is usually accomplished by the electrical discharge or high-intensity light from gas-discharge lamps, such as xenon or krypton gas-discharge lamps. Many other important techniques for excitation exist, however, such as chemical reaction, electron beam preionization, nuclear, and gas dynamic. In the chemical reaction pumped lasers, two or more reactive gases are mixed in a chamber where an exothermic reaction occurs, thus producing the energy required for the population inversion and the lasant as well. In electron beam preionization, a broad beam of electrons is accelerated through a high voltage into a foil from which many more slower electrons are ejected; such electrons then pass through the laser cavity, causing ionization by collision. Excitation is actually achieved by means of an electrical discharge at a voltage too low to sustain ionization but that is more effective at achieving excitation. Nuclear excitation refers to the use of a beam of particles from a nuclear reactor or accelerator to cause excitation. Gas dynamic lasers use differences in decay rates from excited states to achieve a population inversion when a high-temperature gas is forced through supersonic nozzles.

A host medium is one in which the lasant is dispersed. In the ruby laser, Al_2O_3 is the host and serves as a matrix to hold the Cr^{3+} ions. Helium is the host in the HeNe laser and is essential to the process of exciting the lasant, Ne. Some laser types have no host; examples are the semiconductor diode laser and ion lasers like argon and krypton.

5-3 HOW THE LASER WORKS

This section provides a brief qualitative explanation of how lasers work. More detailed information is presented later.

All laser action begins with the establishment of a population inversion by the excitation process. Photons are spontaneously emitted in all directions. Photons traveling through the active medium can *stimulate* excited atoms or molecules to undergo radiative transitions when the photons pass near the atoms or molecules. This factor in itself is unimportant except that the stimulated and stimulating photons are in phase, travel in the same direction, and have the same polarization. This phenomenon provides for the possibility of gain or *amplification*. Only those photons traveling nearly parallel to the axis of the resonator will pass through a substantial portion of the active medium. A percentage of these photons will be fed back (reflected) into the active region, thus ensuring a large buildup of stimulated radiation, much more than the spontaneous radiation at the same frequency. Lasing will continue as long as the population inversion is maintained above some threshold level.

The optical resonator, because of its geometrical configuration, provides for a highly unidirectional output and at the same time, through the feedback process, for sufficient stimulated radiation in the laser to ensure that most transitions are stimulated. The phenomenon of stimulated emission, in turn, produces a highly monochromatic, highly coherent output (some exceptions will be discussed later). The combined action of the resonator and stimulated emission produces an extremely bright light source, even for lasers of relatively low power output.

5-4 TYPES OF LASERS

No attempt will be made to list all the significant laser types commercially available. Most industrial lasers will be briefly described here, with additional detail presented on selected lasers later. The lasers to be discussed at this point are HeNe, ruby, Nd–YAG, Nd–glass, CO_2, dye, excimer, and the semiconductor diode.

The HeNe laser is by far the most common laser in use today. It is a gas laser with He the host and Ne the lasant. The output wavelength is 0.6328 μm (red) for most applications, although it can emit at two ir wavelengths, and power levels range from microwatts to about 50 mW continuous. This is a highly reliable, low-cost laser and is used extensively in such applications as alignment, surveying, ranging, displacement measurement, holography, pattern recognition, communications, surface finish analysis, and flow measurement. The means of excitation in the HeNe laser is a dc glow discharge. The gas atoms that participate in the lasing process are not ionized; so the HeNe laser is a neutral gas laser.

The ruby laser has Al_2O_3 (sapphire) as the host and triply ionized chromium ions, Cr^{3+}, as the lasant. The output wavelength is 0.6943 μm (red). It is operated in pulsed fashion with a low repetition rate because of the poor thermal properties of Al_2O_3. Many joules of energy can be realized in a single pulse, however. This energy has been used for drilling diamonds, sending pulses to the moon, spot welding, hole piercing, and pulsed holography. Excitation is by means of capacitive discharge through flash lamps.

The Nd–YAG and Nd–glass lasers are similar in that both have triply ionized neodymium (knee-oh-dim-e-um) atoms (Nd^{3+}) as the lasant. YAG stands for yttrium–aluminum–garnet, a synthetic crystal also used for simulated diamonds. Its chemical formula is $Y_3Al_5O_{12}$. The output wavelength for both laser types is 1.06 μm (near ir). Nd–YAG lasers are operated continuously or at high repetition rates with average or continuous power output of from a few watts to 1 kW. Applications of Nd–YAG lasers include welding, cutting, hole piercing, and excitation of dye lasers with frequency quadrupling. Nd–glass lasers are pulsed because of the poor thermal characteristics of glass, but extremely high peak powers have been achieved. Lasers of this type are pumped by gas-discharge lamps. Some applications for Nd–glass lasers are welding, cutting, hole piercing, and laser fusion experiments.

There are several different basic CO_2 laser designs, but, in general, they use CO_2 as the lasant and also contain N_2 and He. The emission wavelength is 10.6 μm, which is in the mid-infrared, and the power output is continuous or pulsed with average power ranging from a few watts to tens of kilowatts. Excitation techniques include dc glow discharge, rf, electron beam preionization, chemical, and gas dynamic. As far as industrial lasers are concerned, glow discharge, rf, and electron beam preionization excitation are of greatest interest. Applications for this laser are numerous and include cutting, hole piercing, welding, heat treatment, scribing, and marking. The variety of materials that can be worked is also varied, including paper, plastic, wood, glass, cloth, ceramics, and most metals.

Dye lasers are devices in which the lasant is an organic dye. Because organic dyes fluoresce in a large number of lines (distinct wavelengths), the laser can be tuned to emit at a variety of wavelengths over some wavelength band. By using several dyes, it is possible to obtain thousands of distinct, closely spaced[2] wavelengths in the visible part of the spectrum from a single

---

[2] Line broadening causes the discrete lines to overlap, thus, in effect, producing a continuously tunable laser.

laser with fairly uniform power output at all these wavelengths. Power outputs range from milliwatts to watts. These lasers are pumped by xenon lamps, argon lasers, nitrogen lasers, excimer lasers, or frequency-multiplied Nd–YAG lasers. The output of dye lasers is normally pulsed. Dye lasers are used in spectroscopy, photochemical reaction studies, pollution detection, and surgery.

The excimer laser is rapidly becoming a common industrial processing tool, especially in the semiconductor field. Excimer refers to a class of lasers that produce radiation in the uv region of the spectrum. The term excimer is a contraction of *excited dimer*, which refers to a diatomic molecule in an excited state. Dimer literally means a molecule made up of two similar atoms. The molecules in an excimer laser are actually made up of a halogen and an inert gas atom. Such molecules are only formed when the halogen atom is in an excited state. The bond is very strong, but the excited molecule is very short lived.

When the halogen atom undergoes a transition to the ground state, the bond breaks, releasing a comparatively high energy photon. The transitions can be stimulated, and therefore laser action is achieved.

The following wavelengths are available from excimer lasers: ArF operates at 193 nm, KrF at 248 nm, XeCl at 308 nm, and XeF at 351 nm. A single machine can operate at all these wavelengths by varying the gas mixture and optics. Average power levels of a few hundred watts have been achieved. Because of the short lifetime of the excimer molecules, these lasers have an output that consists of very short pulses, typically on the order of tens of nanoseconds.

The semiconductor diode laser is a distant cousin of the light-emitting diode (LED) in that it is a semiconductor diode that emits light. The diode laser, however, emits coherent radiation with a much narrower wavelength spread than the incoherent emission of an LED. Modern diode lasers are chiefly composed of a variety of ternary compounds,[3] such as $GaAl_xAs_{1-x}$. Feedback in diode lasers is achieved either by cleaving the ends of the diode parallel to each other and perpendicular to the diode junction to form mirror surfaces or by distributed feedback via a diffraction grating etched on top of the chip parallel to the junction. The light is emitted approximately parallel to the junction, rather than roughly at right angles to it as in most LEDs. Physically, diode lasers are small, being on the order of 250 μm on a side by 50 μm thick. These devices, however, can emit peak powers of many watts in 100- to 200-ns pulses with an average power output of several milliwatts. Diodes operating continuously at room temperature can emit several milliwatts of power. Wavelengths vary slightly for $GaAl_xAs_{1-x}$ but are close

---

[3] A compound containing three elements in contrast to a binary compound, which contains two.

to 0.9 μm (near ir). Applications for diode lasers appear to be in fiber-optic communications, pattern recognition, ir illumination, and pollution detection and control.

PROBLEMS

5-1. List six unique properties of laser light.

5-2. List five requirements for lasing action to occur and briefly explain what each means.

5-3. List four important industrial lasers and suggest a possible type of application for each, along with the output wavelength and the range of CW power or energy per pulse.

5-4. Starting with the emission of a spontaneous photon along the axis of a laser, qualitatively describe the lasing action in a CW laser. Use sketches to illustrate your discussion and describe the relationship of the radiation to the initial photon.

5-5. What is the function of the mirrors in a laser?

REFERENCES

LENGYEL, B. A., *Introduction to Laser Physics*. New York: John Wiley & Sons, Inc., 1966.

O'SHEA, D. C., W. R. CALLEN, AND W. T. RHODES, *Introduction to Lasers and Applications*. Reading, MA: Addison-Wesley Publishing Co., 1978.

Chapter 6

Laser Theory

The objectives of this chapter are to provide a sufficient theoretical basis so that readers can pursue additional study on their own and appreciate properly the material presented in later chapters. A basic understanding of what the laser is, how it works, and what its unique properties are will help dispel the mystery that surrounds this useful device. When understood, tools are much more frequently used and used correctly.

Only a brief theoretical treatment can be presented in this limited space. Many results are given without proof, and derivations are kept as simple as possible without loss of meaning.

6-1 POPULATION INVERSION

One requirement for lasing action is a population inversion. Establishing a population inversion requires that more lasant atoms or molecules be in some excited state than in the state to which the most probable transition occurs. This transition must be one in which photons are emitted (radiative transition), as opposed to one in which no photons are emitted (nonradiative transition). In the latter case, the energy ends up as thermal instead of electromagnetic energy.

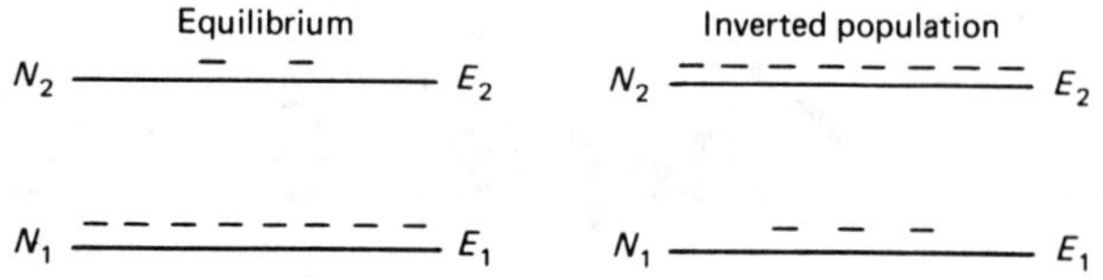

Figure 6-1 Two-level energy-level scheme.

If only two energy levels were involved, the system would be called a two-level system. The only laser that actually resembles this situation is the diode laser[1] since the transitions take place between levels near the band edges of the semiconductor. Most lasers are better approximated by three- or four-level systems. The actual energy level structure of atoms or molecules is much more complicated than that, however. Nevertheless, much can be learned by considering a hypothetical two-level system. Some discussion of three- and four-level lasers will be presented subsequently. With this in mind, we initially discuss population inversion, plus a number of other matters, in the context of a two-level system.

A two-energy-level system is depicted in Fig. 6-1. Here E_1 and E_2 are the energies of the respective levels, and N_1 and N_2 are the number of atoms or molecules per unit volume in each level. The ratio of N_2 to N_1 at thermal equilibrium is given by the Boltzmann equation, which is presented as Eq. (6-1).

$$\frac{N_2}{N_1} = e^{-(E_2 - E_1)/kT} \tag{6-1}$$

Here k is the Boltzmann constant and T is absolute temperature. It can be seen from this relation that N_2 can never exceed N_1 for an equilibrium situation. A population inversion refers to a nonequilibrium situation in which $N_2 > N_1$. It occurs as a result of the excitation process and has been referred to as a negative-temperature situation because Eq. (6-1) can only be made to predict $N_2 > N_1$ if T is negative. This was an attempt to fit a thermal equilibrium result to a nonequilibrium condition and is not considered further here. A population inversion is absolutely necessary for amplification of light to occur. Without a population inversion, a beam of light directed through the medium with photon energy $hf = E_2 - E_1$ will be absorbed.

6-2 STIMULATED EMISSION

Three types of transitions involving photons in laser operation are of interest: absorption, spontaneous emission, and stimulated emission. Absorption occurs when a photon with energy equal to the energy difference between two

---

[1] The ammonia maser is a two-level system.

levels is absorbed, thus causing an excitation. In electronic transitions the electron is raised to a higher energy level; in molecular transitions the vibrational energy of the molecule is increased. Atoms or molecules[2] in excited states eventually drop to a lower energy state spontaneously; in radiative transitions a photon is given off with energy equal to the difference in the energy levels.

The precise theory of the phenomenon of stimulated emission is well beyond the scope of this book, which does not, however, preclude understanding the nature and consequences of the phenomenon. Basically, what happens is that a photon passing sufficiently close to an excited atom causes it to undergo a radiative transition before it would otherwise do so spontaneously. The stimulated photons have the same phase and the same polarization state and travel in the same direction as the stimulating photons. Consequently, gain is achieved when a beam of light of the right frequency (photon energy) is directed through a medium in which a population inversion has been established.

The fact that the stimulators and stimulatees are in phase with each other produces a high degree of both spatial and temporal coherence in lasers. Usually, substantially less than perfect temporal coherence is achieved because of random fluctuations that occur in the laser due to vibrations and thermal variations. Stabilization techniques can be used to increase temporal coherence greatly and hence coherence length. The lateral spatial coherence of laser beams is extremely good.

The occurrence of the phenomenon of stimulated emission is not too surprising with hindsight. The atoms or molecules behave like excited electric dipoles. When these dipoles come under the influence of the electric and magnetic field associated with a photon, the dipole experiences a force. If the phase and frequency of the dipole and photon match (also the spatial orientation of the photon electric field and dipole orientation), this force results in the dipole giving off radiation (a photon). This interaction has often been referred to as "tickling" the excited atoms with the photon to produce the stimulated emission.

6-3 EINSTEIN PROBABILITY COEFFICIENTS

Based on thermodynamic arguments, Einstein established what is essentially the theoretical basis for the laser decades before the first one was built. This theory deals with the probabilities for absorption, spontaneous emission, and stimulated emission of photons. Actually, it is necessary to talk about the rates of the various types of transitions in order to get a grip on the subject. The rate of spontaneous emission can be written $N_2 A_{21}$, where N_2

---

[2] Frequently, the term atom will be used where either atom or molecule could apply.

is the number of atoms per unit volume in the upper state (2) and A_{21} is the probability per unit time that a given atom in an excited state will spontaneously undergo a transition to state 1. The product N_2A_{21} then gives the number of spontaneous transitions per unit volume per unit time (rate per unit volume). The absorption and stimulated rates are a little harder to describe because neither can occur if no radiation field is present. It is reasonable to expect these rates to be proportional to the radiation energy density. Thus the product $N_1\rho_fB_{12}$ gives the rate of absorption of photons per unit volume. The energy density ρ_f is a spectral quantity and therefore has the units J/m^3 Hz. B_{12} is the probability per unit time per unit spectral energy density that a given atom will be excited from state 1 to state 2. The units of B_{12} are complicated because of the absorption rate dependence on ρ_f; they are m^3 Hz/J-s. The rate of stimulated emission can be expressed as $N_2\rho_fB_{21}$, where N_2 and ρ_f are the same as previously introduced. B_{21} has the same interpretation as B_{12} except that it relates to stimulated transitions from state 2 to state 1. At equilibrium, assuming that only optical transitions occur, these rates are related as follows:

$$N_1\rho_fB_{12} = N_2A_{21} + N_2\rho_fB_{21} \tag{6-2}$$

This equation simply states that the rate of upward transitions (absorptions) equals the rate of downward transitions (emissions) at equilibrium for a two-level system. By using the Boltzmann relationship given in Eq. (6-1), the ratio N_1/N_2 can be eliminated from Eq. (6-2), yielding

$$e^{(E_2 - E_1)/kT}\rho_fB_{12} = A_{21} + \rho_fB_{21} \tag{6-3}$$

Setting $E_2 - E_1 = hf$, the energy of a photon absorbed or emitted, and solving for ρ_f yield

$$\rho_f = \frac{A_{21}}{B_{12}e^{hf/kT} - B_{21}} \tag{6-4}$$

Prior to Einstein's work, Planck worked out a model of black-body radiation that correctly predicted the spectral emission and radiation density of a black body, the ideal absorber and radiator. The spectral radiation density at frequency f, according to Planck's theory, is

$$\rho_f = \frac{8\pi hf^3}{c^3} \frac{1}{e^{hf/kT} - 1} \tag{6-5}$$

Because Eqs. (6-4) and (6-5) are equivalent for a two-level system, we can compare them and see what it tells us about the Einstein coefficients. The only way the right sides of Eqs. (6-4) and (6-5) can be equal is if $B_{12} = B_{21}$ and A_{21} is related to B_{21} by

$$A_{21} = \frac{8\pi hf^3}{c^3} B_{21} \tag{6-6}$$

the conclusions reached here are extremely important. The fact that $B_{12} = B_{21}$ means that, for a given radiation density, a stimulated radiation is just as probable for an atom in state 2 as absorption is for an atom in state 1. Furthermore, the coefficient of B_{21} in Eq. (6-6) is numerically 5.6×10^{-16} (SI units) at 10^{14} Hz, which means that in a radiation field of density 10^{-6} J/m³ Hz the rate of spontaneous radiation is 10^{10} times smaller than the rate of stimulated radiation. This result ensures that when a population inversion is established and some feedback mechanism is included to provide reasonable radiation density at the correct frequency, more stimulated emission than spontaneous emission will occur and the device will lase. As noted, all that is needed to achieve amplification is a population inversion and a beam of light of the correct frequency, but it generally must come from a laser.

6-4 AMPLIFICATION

If there were no losses in a medium with an inverted population, the growth in radiation would be geometrical.[3] This situation is illustrated for the usual linear laser or amplifier arrangement in Fig. 6-2. As can be seen, the first

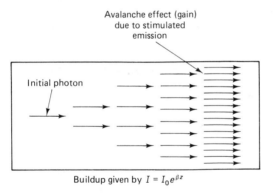

Figure 6-2   Optical amplifier.

photon creates one more; these two, in turn, produce four, which, in turn, produce eight, and so on. A geometrical growth implies that there will be a fixed fractional increase in the number of photons per unit length of the medium. Because the irradiance is proportional to the flux of photons, this can be expressed as

$$\frac{dI}{I} = \beta \, dz$$

where I is irradiance, z distance, and β the fractional increase in irradiance

---

[3] We are no longer talking about an equilibrium situation and it should be recognized that a population inversion in a two-level system cannot be achieved by optical pumping.

per unit length or gain coefficient. Beer's law says that the loss of irradiance in a dielectric medium is given by $dI/I = -\alpha\, dz$, where α is the fraction loss per unit length. Thus the net fractional loss or gain is $dI/I = (\beta - \alpha)\, dz$, which when integrated is

$$I = I_0 e^{(\beta - \alpha)z} \tag{6-7}$$

There are various causes of losses in a laser, such as absorption and scattering in the lasing medium, absorption by windows used to seal discharge tubes, absorption by the mirrors, and, of course, the output. In one complete round trip in a laser of gain length[4] l, the round trip gain, $I/I_0 = G$, is given by $R_1 R_2 e^{(\beta - \alpha')2l}$, where R_1 and R_2 are the reflectances of the mirrors and all other losses are included in α. This can be written

$$G = e^{(\beta - \alpha')2l} \tag{6-8}$$

where

$$e^{-\alpha'2l} = R_1 R_2 e^{-\alpha 2l}$$

Henceforth it will be assumed that all losses are incorporated in α.

It is instructive to derive an expression for the gain coefficient on the basis of the population inversion and the Einstein coefficients. Assuming that spontaneous emission is negligible, we can write the net transition rate as

$$(N_2 \rho_f B_{21} - N_1 \rho_f B_{12})A\,\Delta z = (N_2 - N_1)\rho_f B_{21} A\,\Delta z$$

This transition rate can be equated to the power (irradiance times area) $\Delta I_f A$ generated in the volume element $A\,\Delta z$ divided by the energy per photon hf. Thus

$$(N_2 - N_1)\rho_f B_{21} A\,\Delta z = \frac{\Delta I_f A}{hf} \tag{6-9}$$

In the limit as $\Delta z \to 0$, we then have

$$\frac{dI_f}{dz} = (N_2 - N_1)\rho_f B_{21} hf \tag{6-10}$$

For an unidirectional (linear) beam, $I_f = \rho_f c$, so Eq. (6-10) becomes

$$\frac{dI_f}{dz} = \frac{(N_2 - N_1)I_f B_{21} hf}{c} \tag{6-11}$$

which integrates to

$$I_f = I_0 e^{[(N_2 - N_1)B_{21}hf/c]z} \tag{6-12}$$

---

[4] It is important to distinguish between optical path length between mirrors (see Chapter 1) and gain length, especially in solid-state lasers.

where I_0 is some initial irradiance. Therefore,

$$\beta = \frac{(N_2 - N_1)B_{21}hf}{c}$$

We now substitute for B_{21} in terms of A_{21} and get

$$\beta = \frac{(N_2 - N_1)c^2 A_{21}}{8\pi f^2} \tag{6-13}$$

Recall that A_{21} is the probability per unit time that an excited atom will undergo a spontaneous transition. Therefore, the reciprocal of A_{21} is the average time τ that an atom remains in the excited state before making a spontaneous transition. Now we can write the gain coefficient as

$$\beta = \frac{(N_2 - N_1)c^2}{8\pi f^2 \tau} \tag{6-14}$$

Although we have ignored line-broadening effects (discussed later) and degenerate energy levels (more than one level with the same energy), some interesting conclusions can be drawn from Eq. (6-14).

The most obvious conclusion from Eq. (6-14) is verification of the fact that $N_2 > N_1$ is required to achieve gain, if $N_2 < N_1$ loss is incurred. A second interesting point is that the gain is inversely proportional to the frequency squared (line-broadening effects make this even worse). This statement says that lower gains will be achieved at higher frequencies, and therefore high-frequency lasers will probably be harder to develop than low-frequency ones. It appears to be borne out by the low number of lasers in the uv compared with visible and ir lasers and the total absence of lasers in the x-ray[5] or γ-ray region. The third very interesting point is that the gain is inversely proportional to the spontaneous lifetime τ. It is often pointed out that the upper lasing level should be metastable or a long-lifetime state. It can be seen from Eq. (6-14) that a short lifetime is desirable for high gain. Because a finite amount of time is required to achieve a population inversion, a compromise must be reached. We would like τ as short as possible to provide high gain but sufficiently long to allow the population inversion to be established and maintained in the case of continuous output (CW) lasers. It is also desirable that the lifetime of the upper lasing level be substantially longer than the lifetime of the lower level, although it is not strictly necessary in pulsed lasers.

6-5 POWER OUTPUT FOR CW OPERATION

It is helpful for an understanding of both CW and pulsed laser operation to look at how CW operation relates to the gain coefficient and optical losses.

---

[5] Lasing in the soft x-ray region has been achieved experimentally.

If we assume some initial power P_0 directed along the axis of the laser, then the power flux at the same point after one complete round trip would be

$$P_0 + P_0 e^{(\beta-\alpha)2l}$$

where l is the gain length. The power output of the laser would be

$$P = TP_0[1 + e^{(\beta-\alpha)2l}]$$

where $T = 1 - R$ is the transmittance of the output mirror. The TP_0 term is retained because this power is always being generated spontaneously. After another round trip we have

$$P = TP_0[1 + e^{(\beta-\alpha)2l} + e^{2(\beta-\alpha)2l}]$$

Here it is seen that each existing term is amplified and a new TP_0 is generated. The power expression for a continuously operating laser becomes an infinite series:

$$P = TP_0 \sum_{n=0}^{\infty} e^{n(\beta-\alpha)2l} \tag{6-15}$$

In the case where $n(\beta - \alpha)2l$ is less than zero (which it must be for stable CW operation), this series can be summed and Eq. (6-15) becomes

$$P = TP_0[1 - e^{(\beta-\alpha)2l}]^{-1} \tag{6-16}$$

This result implies that β must be at least slightly less than α for CW laser action, which seems contrary to common sense. However, it must be recognized that if $\beta > \alpha$ then P would increase indefinitely, which is impractical for CW operation or any other kind for that matter. To have equilibrium, the optical power generated must be exactly balanced by optical losses, including the useful output. The gain coefficient β must be slightly less than α because of the small amount of spontaneously generated optical power. Stated another way, the apparent deficit due to the fact that the gain is less than the loss is compensated for by spontaneously produced power. Fortunately, the spontaneous power is negligible compared with the stimulated power in lasers.

The value of β given by $\beta_{th} = \alpha$ is called the *threshold value* of the gain coefficient. The gain must closely approach this value in order for continuous lasing action to occur.

In Eq. (6-8), if the round trip gain, G, is set equal to 1, the threshold gain can be easily obtained

$$\beta_{th} = \alpha - \frac{1}{2l} \ln (R_1 R_2) \tag{6-17}$$

6-6 PULSED OPERATION

In pulsed operation of lasers the gain is raised substantially above the threshold value, generally by means of flash lamps in solid lasers or pulsed electrical discharge in gas lasers. As threshold is exceeded, lasing action begins, but the population inversion continues to increase for some time, resulting in a rapidly increasing power output. This situation quickly leads to a diminution of the population inversion and a drop in power. The power output of a pulsed ruby and Nd–glass laser looks somewhat as shown in Fig. 6-3(a). The power spiking is the result of the periodic buildup and depletion of the population inversion. The gradual decline in peak power is the result of the diminishing power output of the flash lamps. Effectively, the laser is alternately lasing and not lasing. The reason is clear from Eq. (6-14); β is alternating between values much larger than α and less than α as the population inversion alternately increases due to the optical pumping and decreases due to the rapid growth in stimulated emission.

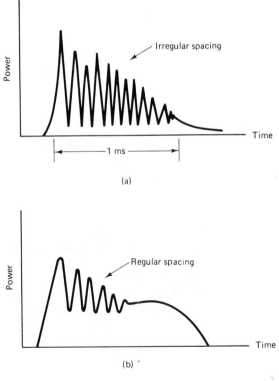

Figure 6-3 Laser pulses.

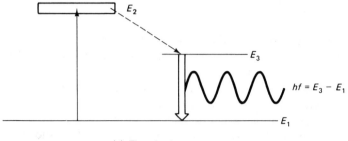

(a) Three-level system

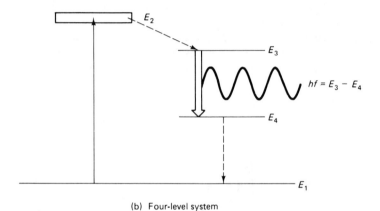

(b) Four-level system

Figure 6-4 Three- and four-level energy-level schemes.

    The spiking is irregular in spacing due to mode hopping. Figure 6-3(b) is representative of Nd–YAG laser pulses if the cavity is not optimized to prevent the quasi-continuous tail that follows the initial spiking. The quasi-continuous portion of the pulse is undesirable for drilling and cutting applications, due to the fact that the peak power is not high enough to produce clean material removal, but is high enough to cause melting. To get around this problem, the cavity must be tuned for single-mode operation. This is generally done by placing an etalon in the resonator, possibly as the output mirror. An *etalon* is a dielectric slab in which reflections occur from both sides. Consequently, internal standing waves are set up inside the etalon. Since the thickness is small, $\delta f = C/(2l)$ is large. The laser can only operate in[6] modes that satisfy both resonators. This reduces the oscillations to one or very few modes. The end result is a spiky output similar to ruby or Nd–glass lasers.

---

[6] See Section 6-9 for a discussion of laser resonator modes.

6-7 THREE- AND FOUR-LEVEL SYSTEMS

In reality, lasers are more nearly approximated by three- or four-level systems, such as those schematically represented in Fig. 6-4.

Two common types of lasers are accurately represented as three-level lasers: ruby and dye lasers. In both cases, excitation is achieved by optical pumping to some level, or band of levels, E_2, followed by rapid nonradiative decay to some lower level E_3. Lasing action then takes place between level E_3 and the ground state E_1. The threshold-level pump energy is very high for these lasers because the entire population of lasant species must be inverted.

In four-level lasers, such as Nd–YAG, HeNe, ion, and CO_2, excitation is to a level or band of levels E_2, followed by rapid nonradiative transition to level E_3, with lasing occurring between that level and another intermediate level E_4. It is hoped that rapid relaxation by nonradiative transition to the ground state E_1 occurs. The potential for CW operation in four-level lasers is much higher than in three-level lasers, for the threshold pump energy is drastically lower: there is no need to invert the entire population of lasant species. It is only necessary that $N_3 - N_4$ exceed some threshold value. Excitation is by optical methods in solid lasers and usually by electrical discharge in gas lasers. In CO_2, HeNe, and metal vapor lasers, various types of energy transfer by collision with other ions, molecules, or atoms play a major role in producing the population inversion.

6-8 RATE EQUATIONS

Primarily for better understanding rather than practical use, the rate equations for three- and four-level lasers are presented here and discussed briefly.

The appropriate equations for the three-level system are

$$N_1 + N_3 = N_1 \tag{6-18}$$

$$\frac{dN_3}{dt} = W_p N_1 - Bp(N_3 - N_1) - \frac{N_3}{\tau} \tag{6-19}$$

$$\frac{dp}{dt} = VB_p(N_3 - N_1) - \frac{p}{\tau_c} \tag{6-20}$$

where $W_p N_1$ accounts for the pumping rate, τ is the lifetime of the lasing state, τ_c is the cavity lifetime of a photon, $B = B_{12} = B_{21}$ is the Einstein stimulated transition coefficient (probability rate per photon),[7] p is the number of photons in the cavity, and V is the mode volume[8] of the active region.

[7] Previously, the Einstein coefficients were based on energy density.

[8] Mode volume is the volume of the active material actually filled by radiation.

Equation (6-18) states that the total number of atoms N_t is shared between the lasing state and the ground state; N_2 is assumed to be negligible because of the rapid decay from pump level. Equation (6-19) expresses the rate of change of atoms in the lasing state as the pump rate minus the stimulated and spontaneous emission rates. Equation (6-20) expresses the rate of change of photons in the cavity in terms of the net stimulated emission rate minus the rate of loss of photons, p/τ_c.

Letting $\Delta N = N_3 - N_1$ represent the population inversion, Eqs. (6-18) to (6-20) can be reduced to the following two equations.

$$\frac{d\,\Delta N}{dt} = W_p(N_t - \Delta N) - 2Bp\,\Delta N - \frac{\Delta N + N_t}{\tau} \qquad (6\text{-}21)$$

$$\frac{dp}{dt} = \left(VB\,\Delta N - \frac{1}{\tau_c} \right)p \qquad (6\text{-}22)$$

Equation (6-22) tells us that lasing action will begin at some critical value of population inversion, $\Delta N_c = 1/(\tau_c VB)$. For $\Delta N > \Delta N_c$, $dp/dt > 0$ and amplification begins. A critical pump rate can be obtained from Eq. (6-21) by setting $d\,\Delta N/dt = 0$, $\Delta N = \Delta N_c$, and $p = 0$. Setting $p = 0$ is justified by the fact that the number of photons present in the cavity prior to the onset of lasing is very small. Thus

$$W_{pc} = \frac{1}{\tau}\frac{N_t + \Delta N_c}{N_t - \Delta N_c} \qquad (6\text{-}23)$$

Equation (6-23) reduces to

$$W_{pc}N_{1c} = \frac{N_{3c}}{\tau} \qquad (6\text{-}24)$$

using $N_t = N_{3c} + N_{1c}$ and $\Delta N_c = N_{3c} - N_{1c}$. This last equation simply states that the rate of pump transitions equals the rate of spontaneous transitions at the critical pump rate for the onset of lasing action. Actually, $N_{1c} \approx N_{3c}$ so that approximately $W_{pc} = 1/\tau$. There must, however, be a large enough population inversion to overcome the losses of the laser.

A steady-state condition for a constant pump rate that exceeds the critical pump rate will be reached where $dp/dt = 0$ and $d\,\Delta N/dt = 0$. This results in

$$\Delta N = \frac{1}{VB\tau_c} = \Delta N_c \qquad (6\text{-}25)$$

$$p = \frac{V\tau_c}{2}[W_p(N_t - \Delta N)] = \frac{N_t + \Delta N}{\tau} \qquad (6\text{-}26)$$

The steady-state population inversion is exactly the same as the critical inversion. Higher pump rates mean proportionately higher numbers of photons in the laser. Moreover, because the power output is directly proportional to the number of photons, the power output is also linearly proportional to the pump rate.

The output power can be calculated from $P_{out} = hfpcT/2L$. This conclusion follows from the fact that each photon has an energy hf and all the energy in the resonator strikes the output mirror in the time $2L/c$, with T giving the fraction emitted. It does not mean, of course, that more and more power output can be achieved by supplying more and more power to the laser. Clearly, the pump rate cannot exceed the rate of return of atoms to the ground state. If fresh atoms or molecules can be supplied to the mode volume, as in some gas and liquid lasers, then physical limitations on flow rates will control the pump rate. Thermal effects due to excessive optical pumping in solids can cause severe damage and thus limit maximum pumping levels. The fraction of power input that produces transitions to the pump band in gas lasers frequently decreases with increasing current after some optimum current level is reached.

It is assumed that $N_1 = N_3 = 0$ for a four-level laser because of the fast transistion rates involved. The critical population inversion is once again given by $\Delta N_c = 1/(VB\tau_c)$ and at steady-state $\Delta N = \Delta N_c$. The critical pump rate W_{pc}, however, equals $\Delta N_c/(N_t\tau)$. In addition, because $\Delta N_c \ll N_t$, then clearly the critical pump rate for four-level lasers is generally much less than for three-level lasers, in fact as much as 100 times less. (It is left as a problem for the reader to write the rate equations for the four-level laser and deduce the results stated in this paragraph by analogy with the three-level laser development.)

6-9 LINE BROADENING

Line broadening refers to the fact that some finite bandwidth must always be associated with any electromagnetic radiation regardless of its source. The radiation from a single isolated atom or molecule has a bandwidth due to what is referred to as natural broadening. *Natural broadening* is a result of the finite lifetime of an atom or molecule in an excited state. The average time that an atom spends in an excited state, prior to dropping to a lower state and giving off a photon, is called the *lifetime* of that state. According to a classical point of view (an incorrect point of view that yields essentially correct quantitative results in this case), the atom is a linear dipole that radiates for a period of time equal to the lifetime τ, thus producing a radiation pulse of length τ in time. It can be shown by Fourier analysis that a single pulse of time length τ must contain a continuum of frequencies with a full width at half-maximum (FWHM) given approximately by

$$\Delta f\tau = 1 \tag{6-27}$$

Thus the shorter the pulse, the greater is the bandwidth. The lifetime, τ, is typically 10^{-8} s. According to a quantum mechanical point of view (the correct point of view), Eq. (6-27) follows from the Heisenberg uncertainty relations.

Other broadening mechanisms become important in liquid, gas, or solid laser media. These mechanisms are referred to as homogeneous and inhomogeneous, terms that characterize the physical processes causing the broadening or, alternatively, as Lorentzian and Gaussian, respectively. The latter names characterize the shape of the lines (power versus frequency curves).

Homogeneous broadening is chiefly caused by collisions. In gaseous media, collisions of the lasant atoms with each other, other species in the mixture, or the walls confining the gas will cause perturbations in the energy (frequency) of the photons emitted. By using the Heisenberg uncertainty principle, the frequency spread can be estimated in terms of the mean time between collisions t_c.

$$\Delta f t_c = 1 \qquad (6\text{-}28)$$

Frequent collisions result in a large bandwidth, and so cooling a system to reduce thermal motion will decrease bandwidth. This phenomenon is referred to as homogeneous broadening because every atom in the gas is subject to the same probability of collision as every other atom. Each atom will have the same statistical distribution of frequencies of emission as all the others.

The mean time between collisions in a gas can be estimated by taking the ratio of the mean free path due to thermal motion, l_{th} (average distance traveled between collisions), to the average thermal velocity, v_{th}. Equations for these quantities are readily obtained from thermodynamic arguments. Thus

$$l_{th} = \frac{2}{\pi} \frac{kT}{pa^2} \qquad (6\text{-}29)$$

$$v_{th} = \frac{kT}{\pi m} \qquad (6\text{-}30)$$

$$t_c = \frac{mkT}{16\pi pa^2} \qquad (6\text{-}31)$$

where m is the molecular mass, p the gas pressure, k the Boltzmann constant, and a the molecular radius.

Homogeneous broadening in crystalline solids is the result of interactions between the lasant species and vibrations in the crystal lattice. Lattice vibrations are quantized and are referred to as *phonons*. So, in effect, interaction of lasant species with lattice vibrations is a particle–particle collision and can be treated just like collisions in a gas if the mean time between collisions, t_c, is known.

Inhomogeneous broadening occurs whenever any type of inhomogeneity causes some lasant species to differ from others in terms of the central frequency of their statistical distribution of emission frequencies. This situation occurs in solids as the result of any defects that make the environments of lasant species vary from point to point.

The most important form of inhomogeneous broadening in gas lasers is due to the Doppler effect. This effect refers to the fact that the frequency of light (as with sound) is a function of the relative speed of the source emitting the radiation and observer. The result of the application of the special theory of relativity to this phenomenon is shown as

$$f = \frac{(1 + v/c)f_0}{\sqrt{1 - v^2/c^2}} \qquad (6\text{-}32)$$

where f_0 is the frequency measured by an observer at rest relative to the emitter, v is the relative speed, and c is the speed of light. The velocity is positive when the emitter and observer are moving toward each other and negative when they are moving away from each other. For lasers, $v \ll c$ so that Eq. (6-32) may be written

$$f = \left(1 + \frac{v}{c}\right)f_0 \qquad (6\text{-}33)$$

The lasant atoms or molecules in a gas laser have, in general, components of their velocity in the direction of the axis of the laser. These velocity components are distributed according to Maxwell's velocity distribution,

$$\frac{dN}{N} = \left(\frac{m}{2\pi kT}\right)^{-3/2} \left[\exp - \frac{m}{2\pi kT}(v_x^2 + v_y^2 + v_z^2)\right]dv_x dv_y dv_z \qquad (6\text{-}34)$$

where dN/N is the fraction of atoms or molecules with velocity between **v** and **v** + **dv,** m is the mass of an atom, and k is the Boltzmann constant. If z is the axis of the laser, integration over the x and y velocity components yields

$$\left(\frac{dN}{N}\right)_z = \left(\frac{m}{2\pi kT}\right)^{1/2} \left[\exp\left(-\frac{m}{2\pi kT}v_z^2\right)\right]dv_z \qquad (6\text{-}35)$$

Clearly, the largest fraction of atoms per unit velocity component parallel to the z axis occurs for $v_z = 0$. As v_z increases, the fraction of atoms per unit velocity decreases rapidly [a plot of $N^{-1}(dN/dv)_z$ versus v_z yields a Gaussian curve]. The frequency emitted by the atoms is a linear function of their velocity; therefore, the line shape will be the same as the velocity distribution. That is, the emitted power as a function of frequency will have a Gaussian shape centered around the frequency emitted for an atom at rest, f_0.

Doppler broadening is a form of inhomogeneous line broadening because each value of v_z represents a distinct group of atoms at any instant, and these atoms can only emit at the frequency specified by Eq. (6-33). It can be shown that the Doppler bandwidth (FWHM) is given by

$$\Delta f = 7.16 \times 10^{-7} \sqrt{\frac{T}{M}} f \qquad (6\text{-}36)$$

where T is absolute temperature and M is the mass in grams of 1 mole of gas.

Figure 6-5 illustrates both homogeneous and inhomogeneous line broadening. For comparative purposes, the bandwidth (FWHM) is assumed to be the same for both. The quantity of $g(f)$ is called a *lineshape factor* and is defined such that $\int_0^\infty g(f)\, df = 1$; if I represents the total irradiance, then $g(f)I = I(f)$ gives the spectral irradiance.

It is now possible to modify Eq. (6-14) for the gain β to take into account line-broadening effects. The gain must be proportional to the lineshape factor, for the more spontaneous power emitted at a given frequency, the greater is the gain at that frequency. Therefore, the spectral gain is given by

$$\beta(f) = (N_2 - N_1) \frac{c^2 g(f)}{8\pi f^2 \tau} \qquad (6\text{-}37)$$

Because the bandwidth is always small compared to f, the variation of f in Eq. (6-36) is negligible and $\beta(f)$ takes the same general shape as $g(f)$. It is worth noting that the maximum value of $g(f)$ is $0.637/\Delta f$ and $0.939/\Delta f$ for the Lorentzian and Gaussian lines, respectively [Svelto and Hanna, 1982]. Thus the maximum gain as a function of population inversion can be estimated if the bandwidth and lifetime τ are known.

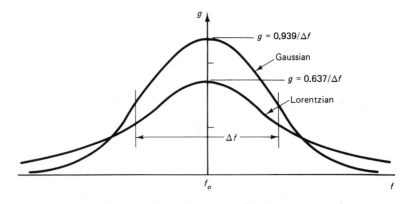

Figure 6-5 Gaussian and Lorentzian line shapes compared.

6-10 Q-SWITCHING, CAVITY DUMPING, AND MODE LOCKING

Techniques for modifying the laser output are discussed here, specifically, techniques for producing, short, high peak power pulses.

Q-switching.   This process involves interrupting the optical resonator in some fashion so that the population inversion process can progress with little or no feedback present to cause lasing. When the population inversion has reached some high level, the resonator interruption is removed and the laser gives out a short pulse with very high peak power. This process is called Q-*switching* in reference to the Q-value of the resonator.

Q-value is defined by

$$Q = \frac{2\pi(\text{energy stored})}{\text{energy lost per cycle}} \tag{6-38}$$

If E is the energy stored in the cavity and τ_c is the lifetime of a photon in the cavity, Eq. (6-38) can be written

$$Q = 2\pi E \left(\frac{E}{f\tau_c}\right)^{-1} = 2\pi f \tau_c \tag{6-39}$$

The Q-value of any type of oscillating system can be shown to be given by the ratio of the oscillator central frequency to the oscillator bandwidth:

$$Q = \frac{f}{\Delta f} \tag{6-40}$$

Hence

$$\Delta f = \frac{1}{2\pi\tau_c} \tag{6-41}$$

which would have been predicted on the basis of the uncertainty principle.[9] This is the ideal resonator linewidth and is much smaller than actual linewidths due to instabilities from vibrations and thermal fluctuations.

Q-value can also be related to cavity losses, particularly mirror reflectances. Assuming that R_1 and R_2 are the mirror reflectances and account for the only optical losses, Eq. (6-42) can be written

$$\frac{E}{\tau_c} = \frac{1 - R_1 R_2 E}{\tau_r} \tag{6-42}$$

where τ_r is the round-trip time for the photon in the resonator. Equation (6-

---

[9] Precisely stated, the uncertainty relation is $\Delta f\, \Delta t = 1/2\pi$.

42) simply expresses the rate of energy loss in two different ways. Thus the average number of round trips for a photon is given by

$$\frac{\tau_c}{\tau_r} = \frac{1}{1 - R_1 R_2} \tag{6-43}$$

If we use Eq. (6-43) and $\tau_r = 2L/c$, Eq. (6-42) then becomes

$$Q = \frac{4\pi f L}{c} (1 - R_1 R_2)^{-1} \tag{6-44}$$

Equation (6-44) illustrates the fact that Q-values for lasers are very high, typically on the order of 1 million.

Techniques for Q-switching have involved rotating mirrors and mechanical choppers for pulses on the order of 0.001 s. Acousto-optical modulators are used for pulses on the order of 10^{-6} s. Pockels and Kerr cells are used for pulses on the order of 10^{-9} s.

Cavity dumping.   Certain types of lasers cannot be Q-switched successfully due to insufficient lifetime in the lasing state; then a procedure called cavity dumping may be used. This process literally allows the lasing energy to build up in the cavity with 100% feedback, and then at some instant the feedback is reduced to a low value and a pulse of high energy is emitted. This technique does not yield the high peak power pulses of Q-switching but is used effectively, particularly in dye lasers.

Mode locking.   Lasers are generally capable of operating in a variety of modes, both axial and transverse. Normally, when a laser is operating in several such modes at once, the individual modes (of different frequencies) are oscillating independently of each other and tend to have totally random phase relations relative to each other. Mode locking is a process in which large numbers of modes are placed in lock step with each other. The results leads to very short (as short as 1 picosecond or less) pulses of extremely high peak power.

A common technique for accomplishing mode locking is through the use of a saturable absorber. The absorber is a material, such as an organic dye, that normally can absorb the radiation in the laser cavity except when all modes are in step, both in phase and in spatial location, and consequently supply more energy to the dye than it can absorb and dissipate in the time of passage of the pulse through the dye. All dye molecules are simultaneously in an excited state. So the dye is saturated and briefly becomes transparent, allowing part of the pulse energy to pass through. The length of the cell containing the dye is chosen to achieve exactly the right condition for saturation.

Once such a mode-locked pulse occurs, part of it is naturally fed back into the laser, and with sufficient gain it saturates the dye each time it passes

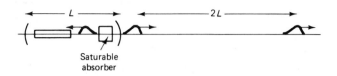

Figure 6-6   Mode-locked pulses.

through it. Thus, once started, the pulse remains in the laser as long as pumping continues. Some laser media are saturable and automatically produce mode locking. Figure 6-6 schematically depicts the mode-locking process.

The distance between pulses is $2L$, the time between the pulses is $\tau_r = 2L/c$, and the bandwidth of the pulses is

$$\Delta f = N \, \delta f \qquad (6\text{-}45)$$

where N is the number of modes locked and δf is the frequency spacing between them. According to the uncertainty principle, the temporal length of the pulse is approximately $(N \, \delta f)^{-1}$. The spatial length (coherence length) is then $c/(N \, \delta f)$. As many as 1000 modes may be locked together; therefore, picosecond length pulses can be obtained. Both pulsed and CW lasers can be mode locked.

PROBLEMS

6-1. (a) Calculate the ratio of the number of atoms in the excited state to the number in the ground state at $T = 300$, 700, and 1000 K if the energy difference is 1 eV.

   (b) What are the wavelength and frequency of the light radiated in a transition from the excited state to the ground state?

6-2. Describe the three types of optical transitions that occur between energy states in a material.

6-3. According to Planck's theory, what is the radiation density of a black body for a frequency of 10^{14} Hz at 1000 K?

6-4. Calculate the ratio of the spontaneous radiation coefficient to the stimulated radiation coefficient at a wavelength of 0.6328 μm.

6-5. If the round-trip gain in a 30-cm-long laser is 1%, calculate the net gain coefficient ($\beta - \alpha$).

6-6. The spontaneous power output for a certain CW laser is 1 μW and the laser power output is 2 mW. If the length of the gain medium is 10 cm, calculate the gain coefficient. Assume that the only loss is due to 2% transmittance at the output mirror.

6-7. Discuss what is meant by three- and four-level lasing systems (illustrate with diagrams) and compare them.

6-8. If the lifetime of a lasing state is 0.5 ms, what is the approximate critical pump rate?

6-9. Repeat the analysis of Section 6-8 for a four-level system.

6-10. Assuming the bulk losses are zero in a CO_2 laser, calculate the threshold gain, given that the gain length is 8 m, and the mirror reflectances are 0.99 and 0.5.

6-11. Estimate the mean time between collisions and the bandwidth due to collision broadening for a molecular gas (CO_2) at 400 K and a pressure of 30 torr.

6-12. Show that the linewidth for a Doppler-broadened line can be written

$$\Delta\lambda = 7.16 \times 10^{-7} \sqrt{\frac{T}{M}}\, \lambda$$

where M is the mass in grams of a mole of gas.

6-13. Calculate the cavity Q-value for a CO_2 laser of 2-m cavity length and reflectances of 100% and 70%.

6-14. (a) What is the average number of round trips for a photon in a laser with mirror reflectance of 100% and 40%?

(b) If the optical cavity length is 1.5 m, what is the cavity lifetime?

6-15. A Nd–glass laser ($\lambda = 1.06$ μm) has an optical cavity length of 1 m and 100 axial modes that are mode locked. What are the frequency bandwidth, spacing between pulses, time between pulses, coherence time, and coherence length of the pulses.

6-16. Calculate the Doppler frequency linewidth for a CO_2 laser if $T = 400$ K.

REFERENCES

DULEY, W. W., *CO₂ Lasers: Effects and Applications*. New York: Academic Press, Inc., 1976.

MAITLAND, A., AND M. H. DUNN, *Laser Physics*. New York: John Wiley & Sons, Inc., 1969.

SIEGMAN, A. E., *An Introduction to Lasers and Masers*. New York: McGraw-Hill Book Co., 1971.

SVELTO, O., AND D. C. HANNA, *Principles of Lasers*. New York: Plenum Press, 1982.

VERDEYEN, J. T., *Laser Electronics*. Englewood Cliffs, NJ: Prentice Hall, 1981.

YARIV, A., *Quantum Electronics*, 3rd. ed. New York: John Wiley & Sons, Inc., 1989.

YOUNG, M., *Optics and Lasers*. New York: Springer-Verlag, 1977.

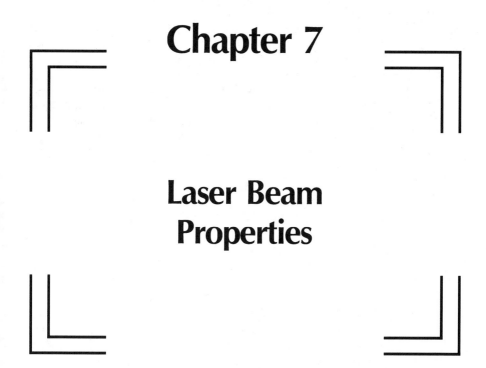

Chapter 7

Laser Beam Properties

This chapter describes the propagation of laser beams and their focusing properties. In addition, the relationship of laser beam parameters, such as beam size and transverse modes, to optical resonator design is covered. Finally, both Gaussian (lowest-order TEM mode) beams and higher-order-mode beams are discussed in connection with the aforementioned properties.

7-1 NATURE OF THE LASER OUTPUT

The beam emitted by a laser consists primarily of stimulated radiation with a small amount of spontaneous radiation (noise) included. Because the radiation is chiefly stimulated, a fairly high degree of temporal coherence occurs. Even in relatively inexpensive HeNe lasers, a coherence length of 1 to 3 m is not unusual. Special stabilizing techniques can be used to increase the coherence length to 100 m fairly easily (Chapter 8). Spatial coherence across the wavefronts in gas lasers is nearly perfect. In solid lasers the transverse spatial coherence may be poor due to a tendency toward lasing in independent filaments within the crystal.

The output of some lasers is linearly polarized. In the case of HeNe lasers, this situation is the result of Brewster windows attached to the end of the discharge tube to contain the gas (Chapter 1). Brewster windows are positioned so that the axis of the laser is at the Brewster angle relative to the normal to the window, as depicted in Fig. 7-1. Light reflected from any surface of one of the Brewster windows is polarized perpendicular to the plane of incidence (plane of the paper in Fig. 7-1). None of the radiation polarized parallel to the plane of incidence is reflected. Therefore, a large loss is introduced for the perpendicular component, resulting in a buildup of the parallel component. Nearly all the radiation in the resonator ends up with parallel polarization and so minimal (ideally zero) reflection losses occur at the windows. What reflection does occur is conveniently reflected *out* of the cavity.

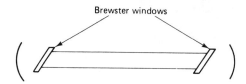

Figure 7-1   Gas laser with Brewster windows.

Some crystalline lasers produce a polarized output because of optical anisotropy of the crystal. If ruby, which has a tetragonal crystal structure, is cut so that its c axis is perpendicular to the cavity, a polarized output parallel to the c axis is produced. If the c axis is parallel to the laser axis, the output is randomly polarized.

A certain degree of linear polarization occurs in lasers where the beam path has been extended by folding the beam with mirrors if the angle between the beam and mirrors is near 45°. This result is clear for dielectric mirrors from the Fresnel equations presented in Chapter 1. Reflection for the perpendicular component is greater than for the parallel component (which is true even though the mirrors may be metal); as a result, this polarization state is preferentially amplified.

In all cases where linearly polarized output is not forced by polarizers, Brewster windows, special crystal properties, or folding mirrors, the output is instantaneously circularly polarized, but the phase fluctuates randomly with time. The fact that the radiation is stimulated results in instantaneous polarization. Thermal and mechanical instabilities cause timewise random changes in the phase of the polarization. Radiation with random polarization produces the same results as unpolarized radiation for most purposes.

It is extremely important to be aware of the nature of the polarization of the laser in many applications. When a laser beam must undergo many reflections in a plane, a large attenuation of the parallel component will occur. Consequently, almost no light is left if the output of the laser is linearly polarized in the parallel sense. More than one holographer has run afoul of this phenomenon. Linear polarization also causes a difference in cutting speed and quality for travel parallel and perpendicular to the polarization direction in high-power laser-cutting applications. Random and circular polarization exhibit no difference.

The beam from a laser expands as though emanating from a point source. In other words, the wavefronts for an axially symmetrical beam are spherical regardless of the TEM mode. The radius of curvature in the near field changes from an infinite value, at the waist, to a finite but linearly increasing quantity as the far field is approached. This situation is shown in Fig. 7-2.

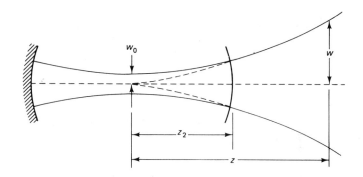

Figure 7-2   Propagating laser beam.

As can be seen, the radius of curvature at the waist is infinite; as the far field is approached, the locus of points defining the beam size approaches an asymptote, and from then on the beam appears to have originated from a point source located at the waist where the asymptotes intersect. As will be seen later, the far field is generally a *long* way from the laser and is rarely reached except in alignment, ranging, and surveying applications.

7-2 OPTICAL RESONATORS

Some form of positive feedback of power is usually necessary if lasing action, in addition to gain, is to occur. This situation can be achieved by the use of distributed feedback, such as diffraction gratings, or by instantaneous feedback with mirrors. Distributed feedback is used with some diode lasers as well as in cases where lasers are integrated into a system, such as in fiber optics or integrated optical circuits. We will discuss only feedback in which mirrors are used. The fraction of power reflected by the output mirror varies from 98% or 99% for low-power lasers to less than 50% for some high-power systems.

Both stable and unstable resonators are used in lasers. A stable resonator is one in which, from a geometrical point of view, certain light rays retrace their paths indefinitely inside the resonator. Conversely, in an unstable resonator the light rays diverge or "walk off" the axis. Most laser resonator designs are stable, although unstable resonators have very useful properties for high-power systems. More detail will be presented later in this chapter on the classification of laser resonators and their effects on laser beam properties. At this point, we will consider the various allowable modes of laser resonators.

If all laser resonators consisted of parallel-plane mirrors (Fabry–Perot interferometer), only axial (longitudinal) modes would need to be considered to determine allowed resonant frequencies. The existence of these modes is based on the electromagnetic boundary condition that only those modes will be supported by the resonator for which the phase change in one round trip, $2L$, is an integral multiple of $360°$ or, expressed mathematically,

$$m\lambda = 2L, \qquad m = 1, 2, 3, \ldots \qquad (7\text{-}1)$$

Be aware that L is the true optical path length of the laser. In Eq. (7-1) λ can be replaced by c/f. Then

$$f = \frac{mc}{2L} \qquad (7\text{-}2)$$

We see from Eq. (7-2) that the frequency spacing between adjacent axial modes is simply $c/2L$.

When one or more spherical mirrors are included in a stable resonator, the wavefronts are spherical, but the mirrors are not necessarily constant phase surfaces. In fact, any mode that is self-reproducing on one complete round trip is an allowable transverse mode. These modes are referred to as *transverse electromagnetic* (TEM_{lp}). They can be circular, but frequently, especially if the laser output is not aperture limited, they take a rectangular form. A few of these modes are depicted in Fig. 7-3.

The subscripts on TEM simply represent the number of nodes in orthogonal directions. Another important mode is the doughnut mode depicted in Fig. 7-4. Actually, this output is the result of a laser operating in the TEM_{01} mode with rapid random variation of the orientation of the node in the plane perpendicular to the z axis (propagation axis).

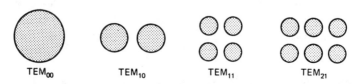

TEM_{00} TEM_{10} TEM_{11} TEM_{21}

Figure 7-3   Transverse-electromagnetic (TEM) mode burn patterns.

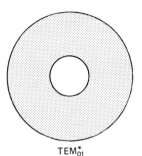

Figure 7-4 Doughnut-mode burn patterns.

TEM$_{01}^*$

The frequency of a given mode, including axial and transverse effects, can be deduced from electromagnetic theory and is given by

$$f_{nlp} = n + (l + p + 1) \frac{\cos^{-1} \sqrt{g_1 g_2}}{\pi} \frac{c}{2L} \tag{7-3}$$

where $g_1 = 1 - L/r_1$, $g_2 = 1 - L/r_2$, and r_1 and r_2 are the radii of curvature of the resonator mirrors. Radii of curvature are taken as positive if measured toward the other mirror and negative otherwise. The frequency spacing between modes due to a difference of 1 in the transverse mode number is

$$\delta f = \frac{\cos^{-1} \sqrt{g_1 g_2}}{\pi} \frac{c}{2L} \tag{7-4}$$

This spacing varies from zero to as much as c/L, which is, in fact, equivalent to $\delta f = 0$.

Unstable laser resonators must satisfy the same axial mode condition as stable resonators. Theoretically, unstable systems can produce a variety of transverse modes; in practice, however, the losses associated with all but the lowest-order transverse mode are so high that unstable resonator lasers tend to operate in the lowest-order transverse mode. This mode is in the form of a ring whose dimensions depend on the resonator design. The ring output is due to geometrical walkoff of the beam, not diffraction, as is implied by the expression *diffraction coupling*, which is often used to describe this type of resonator. A radially symmetrical substructure in the output ring is caused by diffraction. Diffraction effects also cause the output to be less than predicted geometrically. The ring output of an unstable resonator should not be confused with the doughnut mode of stable systems.

7-3 RESONATOR STABILITY AND BEAM PARAMETERS

When referring to the optical stability of a laser resonator, we are talking about whether any rays, geometrically speaking, are trapped within the resonator. We can model the optical cavity as an infinite series of lenses with alternating focal lengths equal to the focal lengths of the mirrors, with spacing between them equal to the optical length of the cavity. If any rays propagating near the axis of such an infinite series of lenses are periodically refocused, the system is stable. If they all diverge, it is unstable. By a purely geometrical analysis, it can be shown that all stable cavities obey the following relation:

$$0 \leq g_1 g_2 \leq 1 \qquad (7\text{-}5)$$

Here $g_1 = 1 - L/r_1$ and $g_2 = 1 - L/r_2$, where r_1 and r_2 are the radii of curvature of the back mirror and the output mirror, respectively, and L is the resonator optical length. These g factors were introduced earlier while discussing the frequencies of TEM modes. A stability diagram is constructed by plotting $g_1 g_2 = 1$ in Fig. 7-5. Points lying on the lines or in the shaded region correspond to stable laser resonators. Points lying on the axes or the hyperbola are referred to as marginally stable because a slight shift in laser length or mirror radius of curvature can cause the system to become unstable.

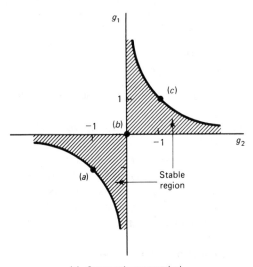

(a) Concentric symmetrical
(b) Confocal symmetrical
(c) Plane mirrors

Figure 7-5 Stability diagram for laser resonators.

Some points of particular interest on the stability diagram are the following:

1. $g_1 = 1$, $g_2 = 1$, which represents the plane parallel mirror, Fabry–Perot interferometer.

2. $g_1 = 0$, $g_2 = 0$, which is a symmetrical confocal resonator. This configuration leads to the lowest diffraction losses in the TEM_{00} mode compared with the higher-TEM modes.

3. $g_1 = -1$, $g_2 = -1$, which is a symmetrical concentric configuration.

A simple rule to determine whether a resonator is stable can be stated. *If either the center of curvature of one mirror or the mirror itself, but not both, falls between the other mirror and its center of curvature, then the resonator is stable* [Yariv, 1968]. If this rule is violated, the resonator is unstable.

Many factors affect the selection of a particular mirror configuration. One mentioned earlier is whether TEM_{00} operation is desired. The confocal resonator exhibits the lowest diffraction losses and the highest selectivity due to a large differential in diffraction losses for the TEM_{00} mode and higher-order modes. This configuration also does a fair job in regard to maximizing mode volume. Mode volume is the volume of the active medium that is actually filled by radiation. Several resonator mirror configurations are shown in Fig. 7-6, along with the approximate mode volume for each.

The confocal mirror configuration is desirable from the standpoint of TEM_{00} mode selectivity, but it is very hard to align because x, y and angular alignment of the mirrors are critical. The concentric configuration has the same problem. Both are marginally stable. The problem of marginal stability is not too difficult because the advantage of the confocal resonator can be nearly achieved by selecting a configuration that lies safely in the stable region of Fig. 7-5, but that is close to the origin.

The plane mirror arrangement has the best mode-volume-filling capability, but results in an output that is heavily multimode, and it is a difficult system to align. Furthermore, the TEM modes are degenerate, making low-order transverse mode operation impossible.

A popular configuration is part (d) in Fig. 7-6 because of the relative ease of alignment and good mode-filling characteristics. This arrangement is frequently used in high-power lasers, with the radius r_1 being several times the optical length of the laser.

Whether a stable resonator laser operates in the lowest-order mode or higher order largely depends on the size of the effective aperture in relation to the cavity length. A gas laser with a long, narrow-bore tubular arrangement will tend to operate low order because of the attenuation of the higher-

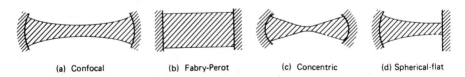

(a) Confocal   (b) Fabry-Perot   (c) Concentric   (d) Spherical-flat

Figure 7-6   Stable resonator configurations.

order modes by the inner walls of the tube. Remember, the higher-order modes are associated with higher diffraction or divergence angles and so are more strongly attenuated by a limiting aperture.

A measure of the tendency for a stable laser cavity to operate low or higher order is the Fresnel number, given by

$$N = \frac{a^2}{\lambda L} \tag{7-6}$$

where a, the effective aperture radius, should be taken as the radius of the smallest aperture in the system if it is not too far from one of the mirrors. In a system where the only limitation is due to the mirrors, assuming that they have equal radii (otherwise, let $a^2 = a_1 a_2$, where a_1 and a_2 are the radii of the two cavity mirrors), the Fresnel number represents the number of Fresnel zones (interference rings) intercepted by one of the mirrors if a uniform plane wave illuminates the other mirror. A Fresnel number of 1 or less generally results in lowest-order mode (TEM$_{00}$) operation.

Stable cavity parameters.   Expressions for the waist size and its location relative to the laser mirrors have been derived for the lowest-order Gaussian and higher-order mode cases [Kogelnik and Li, 1966]. For the lowest-order Gaussian, the waist radius formed inside (or outside) the laser cavity is given by

$$w_0 = \left(\frac{\lambda L}{\pi}\right)^{1/2} \frac{[g_1 g_2 (1 - g_1 g_2)]^{1/4}}{(g_1 + g_2 - 2g_1 g_2)^{1/2}} \tag{7-7}$$

where w_0 is the radius measured to the one over e squared irradiance points. Notice that Eq. (7-7) is an indeterminate expression for certain values of g_1 and g_2. When $g_1 = g_2 = 1$ (two plane mirrors), the spot size is aperture limited and has nothing to do with the resonator parameters. For the $g_1 = g_2 = 0$ case (symmetrical confocal resonator), if g_1 and g_2 are replaced by g and some rearranging is carried out, Eq. (7-7) reduces to

$$w_0 = \left(\frac{\lambda L}{2\pi}\right)^{1/2} \tag{7-8}$$

The location of the beam waist relative to the output mirror, determined by purely geometrical means, is given by

$$z_2 = \frac{g_1(1 - g_2)L}{g_1 + g_2 - 2g_1 g_2} \tag{7-9}$$

Waist location has no meaning for the Fabry–Perot arrangement. The waist is located midway between the mirrors for all symmetrical resonators. For lasers with a plane output mirror and spherical back mirror, the waist is located at $z_2 = 0$ or right at the output mirror.

A useful parameter for characterizing laser beams and simplifying many equations is the Rayleigh range given by

$$Z_R = L \frac{[g_1 g_2 (1 - g_1 g_2)]^{1/2}}{g_1 + g_2 - 2g_1 g_2} \tag{7-10}$$

Thus Eq. (7-7) may be written

$$w_0 = \left(\frac{\lambda Z_R}{\pi} \right)^{1/2} \tag{7-11}$$

or, conversely,

$$Z_R = \frac{\pi w_0^2}{\lambda} \tag{7-12}$$

Although the Rayleigh range can be expressed as a function of minimum spot size, it is a purely geometrical quantity, as indicated by Eq. (7-10). Rayleigh range, therefore, does not depend on the beam mode.

Physically, the Rayleigh range is the distance from the minimum spot (point of infinite radius of curvature of the phase front) at which the radius of curvature reaches a minimum. The phase front radius of curvature is given by

$$R = z \left[1 + \left(\frac{Z_R}{z} \right)^2 \right] \tag{7-13}$$

At $z = Z_R$, $R = 2Z_R$.

7-4   BEAM PROPAGATION FOR STABLE LASER RESONATORS

The output of a stable laser resonator is usually a low-order transverse mode, frequently TEM_{00}, or some higher-order TEM_{lm} mode.[1] It is possible for the output to be multimode, that is, a mixture of various TEM modes. Several axial modes may be present, but we can ignore them as far as beam propagation is concerned.

Lowest-order mode beam propagation.   Solution of the electromagnetic wave equation in rectangular coordinates leads to a particularly useful equation for describing beam propagation. Equation (7-14) gives the irradiance as a function of the cross-sectional coordinates x and y and the distance z from a point where the phase front is flat.

---

[1] Resonator stability is discussed in Section 7-5.

$$I(x,y,z) = I_0 \frac{\sigma_x(0)\sigma_y(0)}{\sigma_x(z)\sigma_y(z)} H_m^2 \left(\frac{x}{\sigma_x(z)}\right) H_n^2 \left(\frac{y}{\sigma_y(z)}\right)$$

$$\times \exp - \left[\frac{x^2}{\sigma_x(z)^2} + \frac{y^2}{\sigma_y(z)^2}\right] \quad (7\text{-}14)$$

The quantities σ_x and σ_y are proportional to the standard deviations of the Gaussian amplitude functions in the respective directions. I_0 is the irradiance at the center of the spot at $z = 0$, which is a point of minimum spot size, usually called the waist, and is normally (but not necessarily) located inside the cavity. H_m and H_n are Hermite polynomials, of which the first few are given here.

$$H_0(q) = 1, \qquad\qquad H_1(q) = 2q$$
$$H_2(q) = 4q^2 - 2, \qquad H_3(q) = 8q^3 - 12q \quad (7\text{-}15)$$

Two useful relationships for Hermite polynomials are presented for future reference.

$$H_n' = 2nH_{n-1}$$
$$H_{n+1} = 2qH_{n-1} - 2nH_{n-1} \quad (7\text{-}16)$$

Here the prime refers to differentiation with respect to the argument. The generating function for Hermite polynomials is given as Eq. (7-17).

$$H_n = (-1)^n e^{q^2} \frac{\partial n}{\partial q^n} e^{-q^2} \quad (7\text{-}17)$$

Referring to Eq. (7-14), we see that the irradiance for the TEM$_{00}$ mode at $z = 0$ is given by

$$I = I_0 \exp - \left(\frac{x^2}{\sigma_x^2} + \frac{y^2}{\sigma_y^2}\right) \quad (7\text{-}18)$$

This may be written

$$I = I_0 e^{-r^2/\sigma^2} \quad (7\text{-}19)$$

for a symmetrical beam.

The definition of spot radius may be taken to be σ, which represents the point at which the irradiance is one over e of its value at the center. A more practical definition of radius for TEM$_{00}$ beams results when Eq. (7-19) is written

$$I = I_0 e^{-2r^2/w^2} \quad (7\text{-}20)$$

Then w is the point at which the irradiance has dropped to one over e^2 of its value at the center. This is the point at which the electric field amplitude has reached one over e of the central value. Over 86% of the total power is contained in a spot of diameter $2w$. Integration of the irradiance over a spot of diameter $2w$ (left as an exercise) yields the following relation between I_0 and P_t, the total power.

$$I_0 = \frac{2P_t}{\pi w^2} \tag{7-21}$$

Figure 7-7 is a plot of irradiance for a TEM_{00} beam.

The manner in which $w(z)$ varies with z has been determined by solving the scalar wave equation for a Gaussian beam and is presented here along with the equation for the phase-front radius of curvature as Eq. (7-22) [Kogelnik and Li, 1966].

$$w(z) = w_0 \left[1 + \left(\frac{\lambda z}{\pi w_0^2} \right)^2 \right]^{1/2} = w_0 \left[1 + \left(\frac{z}{z_R} \right)^2 \right]^{1/2} \tag{7-22}$$

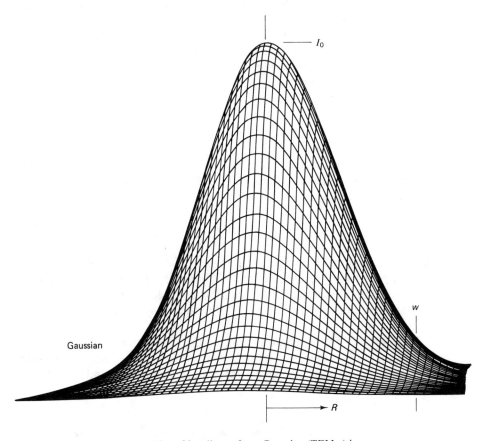

Figure 7-7   Plot of irradiance for a Gaussian (TEM_{00}) beam.

Here w_0 is the spot radius at $z = 0$ and $z_R = \pi w_0^2/\lambda$ is the Rayleigh range. Notice that $w(z)$ is symmetrical about $z = 0$. For very large values of z,

$$\left(\frac{\lambda z}{\pi w_0^2}\right)^2 \gg 1$$

the divergence angle is given by $w(z)/z = \theta = \lambda/\pi w_0$. This is called the *far-field divergence* (diffraction) *angle*. It should be noted that this diffraction angle is 0.52 times the far-field diffraction angle of the central maximum for a plane wave passed through an aperture of diameter $2w_0$. Thus the fundamental (lowest-order Gaussian) beam propagates with about half the divergence of a plane wave of the same initial diameter in spite of the fact that the Gaussian beam is a spherical wave. Also, no higher-order diffraction maxima are associated with the Gaussian beam as with a plane wave incident on a finite aperture.

Equation (7-22) applies equally to a Gaussian laser beam emanating directly from the laser,[2] where z is then measured from the waist produced by the laser optical resonator, or to a focused laser beam. In the case of a focused beam, z is measured from the point of best focus where the phase front is planar. If the divergence angle of the beam entering a lens is small, the beam is focused approximately at the secondary focal point. Because the convergence is rapid (small focused spot size), the lens is in the far field relative to the focused spot. Therefore,

$$w_0 = \frac{\lambda f}{\pi w_l} \qquad (7\text{-}23)$$

where w_0 is the focused spot size and w_l is the spot size at the lens.

The depth of focus of a focused Gaussian beam may also be deduced from Eq. (7-22). Let $w(z) = \rho w_0$ at $z = \pm d$. Then

$$d = \pm \frac{\pi w_0^2}{\lambda} \sqrt{\rho^2 - 1} = \pm \sqrt{\rho^2 - 1} \, z_R \qquad (7\text{-}24)$$

where ρ is a measure of the amount by which the spot radius has increased at a distance d from the focused spot. The depth of focus for a Gaussian beam is substantially greater than for a higher-order or plane wave of the same diameter. For $\rho = 1.05$, Eq. (7-24) reduces approximately to $d = \pm w_0^2/\lambda$.

The depth-of-focus relation given in Eq. (7-24) is acceptable for applications that only require a certain spot size range. In materials processing applications, however, it should be recognized that power and speed also affect depth of focus.

---

[2] If the output mirror is not flat, the lensing effect of the mirror should be taken into account.

Higher-order modes.   Rectangular Hermite–Gaussian higher-order modes are discussed for two reasons. First, many lasers tend to operate in rectangular modes if they operate higher-order. Second, the essential features of all higher-order mode beam propagation are contained in this discussion, and the mathematics are basically straightforward, most having been worked out long ago for the quantum-mechanical simple harmonic oscillator. Equation (7-14) is the basic description of the irradiance of a rectangular higher-order-mode beam propagating in the z direction, with x and y coordinate axes oriented parallel to the rectangular axes of the beam cross section.

Carter proposed a definition of spot size for higher-order rectangular-mode beams that uses the standard deviation of the Hermite–Gaussian functions [Carter, 1980]. Thus the x and y axes are parallel to the sides of the rectangle defining the spot and the dimensions of the spot are $2\sigma_x(z)_m$ by $2\sigma_y(z)_l$.[3] Using the results of the solution for a quantum-mechanical simple harmonic oscillator, for which Hermite–Gaussian functions give the probability function, Carter was able to show that

$$\sigma_s(z)m = (2m + 1)^{1/2}\sigma_s(z) \qquad (7\text{-}25)$$

where $\sigma_s(z)_m$ is the spot size for a mode of order m, s is either the x or y direction, and $\sigma_s(z)$ is the spot size for the corresponding lowest-order mode. Figure 7-8 contains plots of $H_m^2\ (s/\sigma_s)\ \exp\ (-s^2/\sigma_s^2)$ for several values of m. Plots 7-8(a) through (c) represent the irradiance variations in either the x or y directions. Figure 7-8d is a three-dimensional plot of a TEM_{21} mode.

Notice in the cases shown in Fig. 7-8 that σ_{sm} falls outside the outermost irradiance peak. In fact, it may be argued that this is true for all values of m [Luxon and Parker, 1981]. The expression $H_m^2 \exp\ (-s^2/\sigma_s^2)$ in quantum mechanics represents the probability (aside from a constant) of locating a particle undergoing simple harmonic motion at position s relative to its equilibrium position. The quantity σ_{sm} is shown to be the classical amplitude of the oscillator. As m (m is the quantum number) is increased, the correspondence principle of quantum mechanics predicts that the peak probability should occur at a value of s approaching σ_{sm}. At a value of $m = 14$, the peak probability has been calculated to occur at $s = 0.91\sigma_{sm}$. Under no circumstances can the peak probability fall outside the classical amplitude. Therefore, it may be concluded by analogy that all irradiance peaks for Hermite–Gaussian beams are contained within a spot of dimensions $2\sigma_{xl}$ by $2\sigma_{ym}$. Most energy in the beam falls within a spot of this size even for

---

[3] $\sigma_x(z)_l$ is $\sqrt{2}$ times the standard deviation of the function

$$H_m^2 \left(\frac{x}{\sigma_x}\right) e^{(-x^2/\sigma_x^2)}$$

and similarly for $\sigma_y(z)_l$.

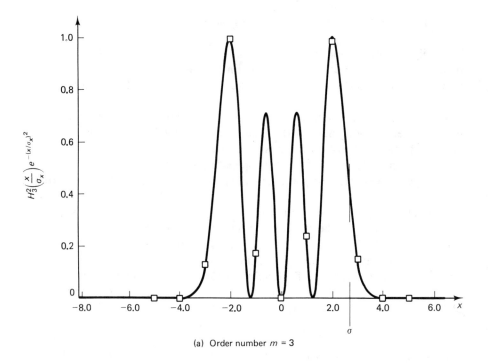

(a) Order number $m = 3$

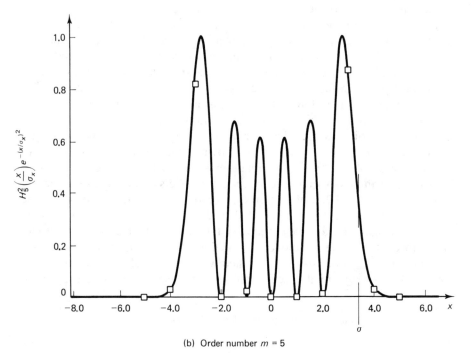

(b) Order number $m = 5$

Figure 7-8 (a–c) Plots of $H_m^2(x/\sigma_x)e^{-(x/\sigma_x)^2}$ with $\sigma_x = 1$, maximum value $= 1$. (d) Three dimensional plot of a TEM$_{21}$ mode.

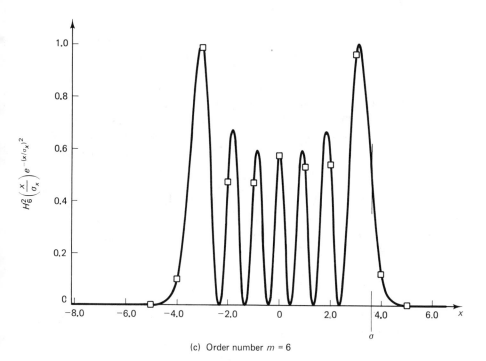

(c) Order number $m = 6$

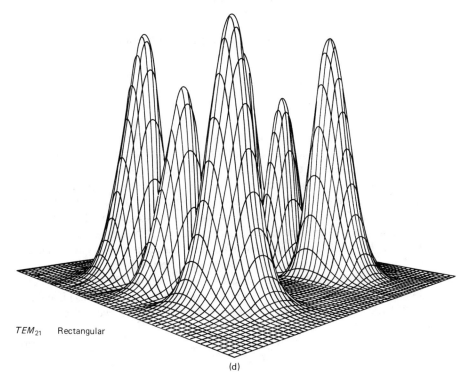

TEM_{21} Rectangular

(d)

Figure 7-8 (*continued*)

relatively low order modes. Numerical integration for a TEM_{11} mode shows that nearly 80% of the beam energy falls within a spot size defined in this way; 85% of the energy falls within such a spot for a TEM_{44}.

It has been shown that the propagation of a higher-order rectangular beam may be written very simply in Eq. (7-22) for the lowest-order beam. This equation is repeated here with the beam size described in terms of σ, since the one over e^2 point has no significance for a higher-order beam.

$$\sigma_s(z) = \sigma_{s0} \left[1 + \left(\frac{\lambda z}{2\pi\sigma_{s0}^2} \right)^2 \right]^{1/2} \tag{7-26}$$

In. Eq. (7-26), σ_s is the spot size for s equal to either x or y, and σ_{s0} is the spot size for the focused beam. The modified equation for the spot size $\sigma_s(z)_m$ of the mth higher-order mode is

$$\sigma_s(z)_m = \sigma_{s0}(2m + 1)^{1/2} \left[1 + \left(\frac{\lambda z}{2\pi\sigma_{s0}^2} \right)^2 \right]^{1/2} \tag{7-27}$$

and $\sigma_{s0}(2m + 1)^{1/2} = \sigma_{s0m}$, which is the spot size for the focused beam. The far-field divergence angle is given by

$$\theta = \frac{\sigma_s(z)_m}{z} = \frac{(2m + 1)^{1/2}\lambda}{2\pi\sigma_{s0}} \tag{7-28}$$

It is instructive to express the far-field divergence angle in terms of the actual focused spot size, σ_{s0m}.

$$\theta = \frac{(2m + 1)\lambda}{2\pi\sigma_{s0m}} \tag{7-29}$$

Using Eq. (7-29), it is clear that the divergence angle for equal spot size is actually $(2m + 1)$ times larger for the higher-order beam.

The quantity σ_{sm} is a logical mathematical definition of spot size because all the irradiance peaks and most of the power fall within a spot of area $2\sigma_{xm}$ times $2\sigma_{yn}$. This definition, however, does not lend itself to direct measurement under any circumstances. It can be determined indirectly by making power measurements through apertures, a process with many difficulties. There is a more practical definition, at least for lasers where mode patterns can be readily viewed, such as burn patterns in acrylic for CO_2 lasers. This definition simply uses the distance between the centers of the outermost irradiance peaks in both the x and y directions. For convenience, half this distance will be used and symbolized by x. Equation (7-27) is easily modified if the relation between x and $\sigma_s(z)_m$ is known. If we write

$$x_m = \frac{1}{k_m} \sigma_s(z)_m \tag{7-30}$$

then Eq. (7-18) may be written:

$$x_m = x_0(2m - 1)^{1/2} \left[1 + \left(\frac{\lambda z}{2\pi k_m^2 x_0^2} \right)^2 \right]^{1/2} \qquad (7\text{-}31)$$

The values of k_m have been numerically computed for m values up to 14 and are listed in Table 7-1 [Luxon and Parker, 1981].

Equation (7-31) is more useful if it is written entirely in terms of the minimum spot size $x_{0m} = x_0(2m + 1)^{1/2}$.

$$x_m = x_{0m} \left[1 + \left(\frac{\lambda z(2m + 1)}{2\pi k_m^2 x_{0m}^2} \right)^2 \right]^{1/2} \qquad (7\text{-}32)$$

The far-field divergence angle may now be written:

$$\theta = \frac{\lambda(2m + 1)}{2\pi k_m^2 x_{0m}} \qquad (7\text{-}33)$$

When a laser beam is focused with a mirror or lens, the convergence is so rapid that the focusing element is in the far field in terms of distance from the focused spot. Assuming that the best focus occurs, to a sufficient approximation, at the focal point, Eq. (7-32) reduces to Eq. (7-34) for the focused spot size.

$$x_{0m} = \frac{\lambda f(2m + 1)}{2\pi k_m^2 x(f)_m} \qquad (7\text{-}34)$$

It is worth noting the inverse relationship between spot size and divergence angle. This leads to the following important result: when a laser

TABLE 7-1   Values of k_m for $m = 1$
through $m = 14$

m	k_m
1	1.73
2	1.42
3	1.30
4	1.24
5	1.20
6	1.18
7	1.16
8	1.14
9	1.13
10	1.12
11	1.11
12	1.11
13	1.10
14	1.10

beam of any order is expanded and recollimated, the divergence angle decreases proportional to the increase in beam size.

A useful definition of death of focus for higher-order-mode beams can be deduced from Eq. (7-32) in the same manner that Eq. (7-22) was deduced from Eq. (7-21). Again, let $z = d$, $D_s(d)_m = \rho D_{s0m}$, and solve for d:

$$d = \pm 2\pi(\rho^2 - 1)^{1/2} \frac{k_m^2 x_{0m}^2}{(2m + 1)\lambda} \tag{7-35}$$

It is important to note the inverse effect of the mode number on depth of focus. This effect is compensated for, however, by the larger focused spot size associated with high-order beams.

Phillips and Andrews (1983) carried out a similar analysis for circular higher-order modes (Laguerre–Gaussian). Their result is simpler in that only one dimension is required to describe the beam size, the radius r. The radius as a function of distance from a minimum (waist) is given by

$$r = r_0 \left[1 + \left\{ \frac{\lambda z(2p + q + 1)}{\pi r_0^2} \right\}^2 \right]^{1/2} \tag{7-36}$$

In Eq. (7-36), p is the number of circular nodes and q is the number of radial (straight) nodes. Unlike the expression for rectangular modes, Eq. (7-36) reduces to the equation for the TEM_{00} mode when $p = q = 0$. As in the rectangular case, the power contained in a spot of radius r varies with the mode, but it exceeds 90% of the total power for all except the TEM_{00} mode, for which it is 86.5%. Hence this is a practical definition of spot size for circular mode beams.

All the beam equations can be written in a generic fashion by introducing a mode factor M, which has the value $M = (2m - 1)/2k_m^2$ for rectangular modes, $M = (2p + q + 1)$ for circular modes, and $M = 1$ for the TEM_{00} mode. For example, the propagation equation becomes

$$u = u_0 \left[1 + \left(\frac{\lambda z M}{\pi u_0^2} \right)^2 \right]^{1/2} \tag{7-37}$$

where u is w, x or r, whichever best suits the situation.

Based on Carter's results, Eq. (7-7) has also been modified by Luxon and Parker for Hermite–Gaussian higher-order modes [Luxon and Parker, 1981]. By using the definition of spot size introduced earlier for higher-order modes (half the distance between outer irradiance peaks), the waist size becomes

$$x_{0m} = \left[\frac{\lambda L(2m + 1)}{2\pi k_m^2} \right]^{1/2} \frac{[g_1 g_2(1 - g_1 g_2)]^{1/4}}{(g_1 + g_2 - 2g_1 g_2)^{1/2}} \tag{7-38}$$

The relation for the location of the waist is not altered by higher-order mode operation, for this location depends on purely geometrical factors. In generic form, using the Rayleigh range, this equation becomes

$$u_0 = \left(\frac{\lambda Z_R M}{\pi} \right)^{1/2} \tag{7-39}$$

Effective thin-lens equation.   Self has developed a useful equation that resembles the thin-lens equation, which relates Gaussian waist locations to lens focal length and the Rayleigh range [Self, 1983]. A thin lens transforms the radius of curvature of a Gaussian beam according to

$$\frac{1}{R_1} + \frac{1}{R_2} = \frac{1}{f} \tag{7-40}$$

where R_1 is radius of curvature entering the lens and R_2 is radius of curvature leaving the lens. The sign convention is the same as of o, i and f for a thin lens as given in Chapter 1. Using Eq. (7-43) along with Eq. (7-22) and the fact that a thin lens leaves the beam size unchanged immediately on either side of the lens leads to

$$\frac{1}{o + Z_R^2/(o - f)} + \frac{1}{i} = \frac{1}{f} \tag{7-41}$$

where o is the waist location for light entering the lens and i is the waist location for light leaving the lens. The sign convention on o and i is still the same as in Chapter 1.

Self also presented an expression for magnification,

$$m = \left[\left(1 - \frac{o}{f} \right)^2 + \left(\frac{Z_R}{f} \right)^2 \right]^{-1/2} \tag{7-42}$$

which can be used for spot size determination.

It has been pointed out by Luxon, Parker, and Karkheck, 1984, that Eqs. (7-41) and (7-42) are valid for beams of all modes. For Hermite–Gaussian beams, the Rayleigh range is given by Eq. (7-43):

$$Z_R = \frac{2\pi k_m^2 D_{s0}^2}{\lambda(2m + 1)} \tag{7-43}$$

The transformed Rayleigh range is given by $Z_R' = m^2 Z_R$.

7-5 UNSTABLE RESONATORS

Unstable optical resonators are frequently used in high-power, high-gain lasers in which a relatively low percentage of feedback is required to achieve lasing operation. Although an infinite number of possible unstable resonator

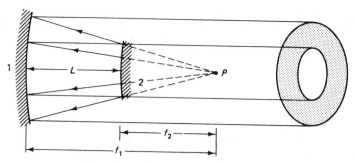

Figure 7-9 Unstable confocal resonator.

configurations is possible, Fig. 7-9 typifies the type used in commercial lasers, particularly multikilowatt CO_2 lasers.

In an unstable cavity the rays "walk off" the axis and exit past the output mirror unless some mirror arrangement is used to intercept the rays and deflect them in some other direction (as is usually the case). In any event, the walk off of the rays leads to a ring-shaped output in the near field with some substructure due to diffraction effects.

Unstable resonators are capable of producing a variety of transverse modes, just as stable resonators do. Yet the losses associated with the higher-order modes are so great that, in practice, such lasers nearly always operate in their lowest-order transverse mode. This mode can result in a divergence angle and focusing capability approximately equal to the Gaussian mode. When focused, the irradiance distribution is similar to that of a focused Gaussian beam. The hole in the center of the beam in the far field is filled in, and the irradiance distribution again is approximately Gaussian.

Some useful results can be deduced from Fig. 7-9 for a confocal unstable resonator. The ratio of a_1 to a_2, the outer and inner radii of the output ring, is called the magnification m and is related by similar triangles to the mirror radii of curvatures by

$$m = \frac{a_1}{a_2} = \frac{r_1/2}{r_2/2} = \frac{r_1}{r_2} \tag{7-44}$$

where use has been made of the fact that the focal length of a spherical mirror is half its radius of curvature. Note that a_2 is the radius of the output mirror, but a_1 is the minimum radius of the back mirror.

If it is assumed that the radiation uniformly fills the back mirror up to a radius a_1, the fraction of the power emitted (geometrical loss G) is given by

$$G = \frac{\pi(a_1^2 - a_2^2)}{\pi a_1^2} = \frac{m^2 - 1}{m^2} = \frac{r_1^2 - r_2^2}{r_1^2} = 1 - \frac{r_2^2}{r_1^2} \tag{7-45}$$

This provides a reasonable estimate of the radiation geometrically coupled out of a confocal unstable resonator. It is also useful to note that

$$L = f_1 - f_2 = \frac{r_1}{2} - \frac{r_2}{2} \tag{7-46}$$

where all quantities are taken to be positive.

An equivalent Fresnel number can be defined for unstable resonators and for the confocal resonator discussed here [Svelto, 1976].

$$N_{eq} = \frac{1}{2} \frac{a_2^2}{\lambda L} (m - 1) \tag{7-47}$$

For m near 1, $N_{eq} < N$ and excellent mode discrimination occurs for all half-integer values of N_{eq}.

PROBLEMS

7-1. The total power emitted by a CW laser operating in the TEM_{00} mode is 10 W. If the focused spot size is 0.2 mm, calculate the irradiance at the spot center.

7-2. The beam radius for a HeNe laser ($\lambda = 0.6328$ μm) operating TEM_{00} is 0.5 mm. Determine the spot size at distances of 1 and 500 m from the waist.

7-3. If the spot radius for a HeNe TEM_{00} beam entering a 12.5-cm focal length lens is 1.5 cm, calculate the focused spot size.

7-4. Calculate the depth of focus of the conditions given in Problem 7-3 for a spot size increase of 10%.

7-5. A CO_2 laser ($\lambda = 10.6$ μm) operates in a rectangular TEM_{64} mode. The beam dimensions, outer peak-to-peak distance, are 2.5 and 2.3 cm. Calculate the focused spot size for both spot dimensions for focusing with a 25-cm focal length lens.

7-6. Calculate the depth of focus for the conditions given in Problem 7-5 for a 10% spot size increase.

7-7. A HeNe laser operating TEM_{00} has a resonator length of 1 m and mirror radii of $r_1 = 3$ m and $r_2 = 2$ m. Locate the waist and determine the spot size at the waist. Also, determine the spot size and divergence angle at the output mirror (r_2).

7-8. A CO_2 laser operating rectangular TEM_{57} has a resonator length of 5 m and mirror radii of $r_1 = 30$ m and $r_2 = 20$ m. Locate the waist and determine spot dimensions at the waist. Also, determine the spot dimensions and divergence at the output mirror (r_2).

7-9. Prove whether the following mirror configurations are stable and plot them on a stability diagram:
(a) $L = 2$ m, $r_1 = r_2 = 2$ m
(b) $L = 1.5$ m, $r_1 = r_2 = 0.75$ m
(c) $L = 5$ m, $r_1 = 30$ m, $r_2 = 20$ m
(d) $L = 1$ m, $r_1 = 4$ m, $r_2 = -2$ m
(e) $L = 1.5$ m, $r_1 = 3$ m, $r_2 = -2$ m
(f) $L = 3$ m, $r_1 = 2$ m, $r_2 = 2.5$ m

7-10. A solid laser consists of mirrors separated by 50 cm with a laser rod 10 cm in length with a refractive index of 1.5 and a Q-switch with a length of 2.5 cm and a refractive index of 1.45. What is the frequency spacing between axial modes for this laser?

7-11. Calculate the frequency spacing between a TEM_{n10} and a TEM_{n20} mode for a laser with an optical cavity length of 5 m and mirror radii of $r_1 = 30$ m and $r_2 = 20$ m.

7-12. Design a stable laser resonator of 30-cm length that will provide a waist at the output mirror with a diameter of 1 mm for a HeNe laser operating TEM_{00}. What should the maximum aperture be for this laser to ensure TEM_{00} operation?

7-13. Design a confocal unstable resonator with a length of 1 m and an output mirror radius of 2.5 cm that will provide 40% output coupling. What are the magnification and equivalent Fresnel number for this resonator? Do you expect good mode discrimination for this resonator if $\lambda = 10.6$ μm?

7-14. For Problem 7-8, treat the output mirror as a thin lens and determine the new waist location, size, and Rayleigh range. (Assume $n = 2.4$ and the outer surface is flat.) Use this information to locate the waist and its size produced by a 12.7-cm focal length lens placed 10 m from the output mirror. Compare your results with the results based on the assumption that the waist produced by the lens occurs in the focal plane.

REFERENCES

CARTER, W. H., "Spot Size and Divergence for Hermite–Gaussian Beams of Any Order," *Applied Optics*, **19**, No. 7, April 1980, 1027–29.

KOGELNIK, H., AND T. LI, "Laser Beams and Resonators," *Applied Optics*, **5**, No. 10, October 1966, 1550–67.

LUXON, J. T., AND D. E. PARKER, "Higher-order CO_2 Laser Beam Spot Size and Depth of Focus Determination," *Applied Optics*, **20**, No. 10, June 1, 1981, 1933–35.

———, "Practical Spot Size Definition for Single Higher-order Rectangular-mode Laser Beams," *Applied Optics*, **20**, No. 11, May 15, 1981, 1728–29.

———, D. E. PARKER, AND J. KARKHECK, "Waist Location and Rayleigh Range for Higher-order Mode Laser Beams," *Applied Optics*, **23**, No. 13, July 1, 1984, 2088–90.

MAITLAND, A., AND M. H. DUNN, *Laser Physics*. New York: John Wiley & Sons, Inc., 1969.

NEMOTO, S., AND T. MAKIMOTO, "Generalized Spot Size for a Higher-order Beam Mode," *Journal of the Optical Society of America*, **69**, No. 4, April 1979, 578–80.

PHILLIPS, R. L., AND L. C. ANDREWS, "Spot Size and Divergence for Laguerre Gaussian Beams of Any Order," *Applied Optics*, **22**, No. 5, March 1, 1983, 643–44.

SELF, S. A., "Focusing of Spherical Gaussian Beams," *Applied Optics*, **22**, No. 5, March 1, 1983, 658–61.

SIEGMAN, A. E., *An Introduction to Lasers and Masers*. New York: McGraw-Hill Book Co., 1971.

SVELTO, O., *Principles of Lasers*. New York: Plenum Press, 1976, 135–42.

VERDEYEN, J. T., *Laser Electronics*. Englewood Cliffs, NJ: Prentice Hall, 1981.

YARIV, A., *Quantum Electronics*. New York: John Wiley & Sons, Inc., 1968, 226.

YOUNG, M., *Optics and Lasers: An Engineering Approach*. New York: Springer-Verlag, 1977.

Chapter 8

Types
of Lasers

Most of the lasers found in industrial applications are discussed in this chapter, including some, such as the dye laser, that are used primarily in laboratory facilities. In most cases, only general descriptions of wavelength, output power, mode of operation, and applications are given. As a specific example, the CO_2 laser is discussed in some detail.

Laser types can be categorized in a number of ways. In general, there are gas, liquid, and solid lasers. Subclassifications of neutral gas, ion, metal vapor, and molecular lasers exist in the category of gas lasers. The only liquid laser discussed here is the dye laser. Solid lasers may be crystalline, amorphous (at least in the case of the Nd–glass laser), or even a semiconductor.

8-1 HeNe LASERS

The HeNe laser is the most common of all the visible output lasers (see Fig. 8-1 for a photograph). It emits in the red part of the visible spectrum at 0.6328 μm. The power levels available range from 0.5- to 50-mW CW output. Most HeNe lasers are constructed with long, narrow-bore (about 2-mm diameter is optimum) tubes, resulting in preferential operation in the TEM_{00}

Figure 8-1   Photograph of a HeNe laser with cover removed. (Courtesy of Spectra-Physics, Inc.)

mode, although TEM^*_{10} may occur if good alignment of the mirrors is not maintained.

The electrical glow discharge is produced by a dc voltage of 1 to 2 kV at about 50 mA. A starting voltage of several kilovolts is applied to get the discharge started. The HeNe laser is a low-gain device that requires optimally 99% feedback to sustain oscillation. Consequently, the mirrors are multilayer dielectric on a dielectric substrate for both back and output mirrors. The output power is 25 mW/m of tube length.

HeNe lasers have a gas mixture with Ne at a partial pressure of 0.1 torr[1] and a total pressure of 1 torr. Ne is the lasant, but He plays an important role in the excitation process. In a four-level process, He is excited by collisions with electrons to a level coincident with an excited state of Ne. During collisions, the energy is resonantly transferred from He to Ne. The Ne decays to one of three possible intermediate states with three possible emission lines, one at 0.6328 μm and two in the ir at 1.15 and 3.39 μm. The mirrors act as selective filters for the 0.6328-μm wavelength.

Figure 8-2 shows the energy-level diagrams of He and Ne and the pertinent transitions. The excited states of He that are involved are metastable, thus ensuring a high probability of resonant energy transfer to Ne. Neon atoms relax to the 1 S level after lasing. This state is metastable and

---

[1] A torr is approximately 1 mm of mercury.

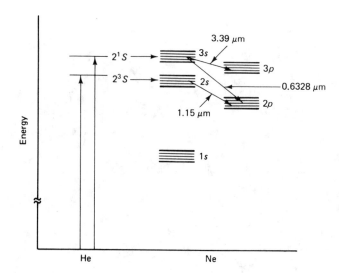

Figure 8-2   Energy-level diagram for HeNe laser.

must be depopulated by collisions with the wall of the tube. Therefore, increasing the bore of the tube past about 2 mm will decrease power output rather than increase it.

Some HeNe lasers are designed to operate in a few low-order modes, or multimode operation. They may also be purchased with either linear or random polarization. The mirrors are attached directly to the tube for random polarization. At least one Brewster window is used for linearly polarized output.

The bandwidth of HeNe lasers is 1500 MHz because of Doppler broadening. Consequently, these lasers generally operate in several axial modes. Shortening the cavity to about 10 cm usually ensures single axial mode operation.

In single axial mode operation, there will be just one dip in the gain curve due to "hole burning." This dip can be centered by adjusting the cavity length. As the laser frequency drifts from this central value, the power will begin to increase (a transient effect). The power change can be monitored by a beam splitter and photodetector. If one mirror is mounted on a piezoelectric crystal, a servosystem can be used to stretch or contract the crystal slightly (by means of an applied voltage), thus adjusting the cavity length continuously to maintain minimum power. In this manner, a frequency stability of one part in 10^9 can be achieved.

High stability can be gained by placing a gas in a separate cell with a coincident absorption line in the laser cavity. I_2^{129} is used with HeNe lasers. I_2^{129} has a very narrow absorption line at 0.633 μm. By locking the HeNe laser on the center of this absorption and saturating the I_2^{129}, frequency stability of one part in 10^{12} to 10^{13} can be achieved.

The diversity of HeNe laser applications is too great to cover exhaustively and so some representative applications are mentioned here. More detail on selected low-power applications can be found in Chapters 9 and 10.

HeNe lasers are used in bar code reading, image and pattern recognition, alignment operations in industry, construction, sewer, and tunnel work, and surveying. In addition, they have a variety of nondestructive testing (NDT) applications, including surface flaw and roughness determination. Measurement applications include interferometric displacement measurement, absolute size measurement, and edge location, using various types of detectors and detector arrays. Frequently, HeNe lasers are used to align other lasers and to locate the point of incidence on a workpiece for a high-power laser.

HeNe lasers are also used in line-of-sight communications applications, such as control of construction equipment like earth graders to maintain proper grade or depth requirements. A major application of HeNe lasers is in the recording and playing back of holograms.

8-2 ION LASERS

Common ion lasers are argon (Ar), krypton (Kr), and xenon (Xe). The power levels of these lasers vary from milliwatts to several watts. The major wavelengths available for each are listed in Table 8-1.

The ion lasers, as the name implies, use an ionized gas as the lasant in an electrical discharge in a plasma. Basically, they are four-level systems, but it is the excited states of the ions that are involved in the lasing process. Excitation is a two-collision process: the first collision ionizes the atom; the second provides the necessary excitation.

The current densities required in ion lasers are about 1000 A/cm², which limits the tube bore diameter to a few millimeters. The ions drift toward the

TABLE 8-1   Wavelengths and Power Ranges for Ion Lasers

Laser	Wavelength (μm)	Power Range (W)
Ar	0.5145 $\rbrace$ 0.488	0.005–20
Kr	0.6471	0.005–6
Xe	0.995–0.5395	200 pulsed

cathode, and so a return path outside the discharge is provided, usually by means of a segmented BeO tube. An axial magnetic field is applied by means of a current-carrying coil wrapped around the tube. This step tends to prevent electrons from losing energy through collisions with the wall since the magnetic field causes them to undergo a spiral motion along the tube axis.

Argon lasers are used extensively in surgery, particularly of the eye. This device is used to bounce light off the retroreflectors that the astronauts placed on the moon for moonquake and meteor impact studies, as well as measurement of the distance between the earth and moon. The Ar laser, along with the Kr laser, is used in spectroscopic work, such as Raman spectroscopy. Both lasers are tunable over several wavelengths in addition to the ones listed in Table 8-1. The Xe laser can be used for some materials processing, such as cutting or removal of thin metallic films. In this type of work the Xe laser is operated in pulsed fashion with peak powers of around 200 W.

The Ar laser has been used in holographic work to achieve shorter exposure times and thereby minimize the effects of vibration. The development of high-speed, high-resolution emulsions for holography has reduced the need for Ar lasers in holographic work.

8-3 METAL VAPOR LASERS

Many types of metal vapor lasers exist, but the He–Cd and He–Se lasers are the most important commercial ones. Although essentially ion lasers, they behave similarly to the HeNe laser in that the metal ions are excited by collisions with excited He atoms. The excitation process in the He–Cd laser is called *Penning ionization*. In this process, an excited metastable He atom collides with a Cd atom, thereby ionizing it and raising it to an excited ionic state. Lasing action takes place between the excited ionic state and the ionic ground state of Cd. The He–Cd laser has wavelengths at 0.441 μm in the blue and 0.325 μm in the uv. Excitation in the He–Se laser is by charge transfer. An excited, ionized, metastable He atom collides with a neutral Se atom, transfers the charge to it, and excites it. Emission occurs between excited states of the Se ion, with about 19 lines possible in the visible spectrum.

In metal-vapor lasers the metal is vaporized near the anode and condenses at the cathode, a process referred to as *cataphoresis*. Consequently, the metal is consumed, for example, at a rate of 1 g of Cd per 1000 h of operation.

The power output of these metal-vapor lasers is in the 50- to 100-mW range. The current densities required for the one-step excitation process in metal-vapor lasers are much lower than for single-element ion lasers, and so they can be air cooled and operated from 110-V ac.

Some applications for metal-vapor lasers are light shows, full color image generation, spectroscopy, and photochemical studies.

8-4 DYE LASERS

Dye lasers, as the name suggests, use an organic dye as the lasing medium. Commercial dye lasers are pumped by nitrogen, argon, Nd–YAG, and excimer lasers, although Xe lamps can be used. Operation is usually pulsed at 100 pps, but average powers of 10 mW are attained.

The organic dyes used, such as rhodamine 6G (xanthene dye), have hundreds of overlapping spectral lines at which they can lase. Consequently, a dye laser with a specific dye can, in effect, be tuned continuously over a large portion of the visible part of the spectrum. By using several dyes, which are automatically interchanged, the entire visible portion of the spectrum, as well as some of the ir and uv, can be scanned with relatively equal power output at each wavelength. This capability is of great value in certain types of spectroscopic investigations in or near the visible wavelengths and also in photochemistry, including uranium isotope separation and pollution detection.

Figure 8-3 is a representative energy-level diagram for an organic dye. The diagram is composed of two types of states: the singlet states labeled

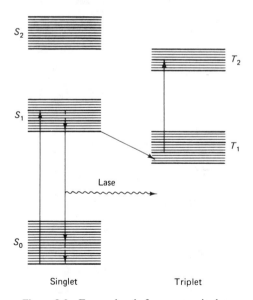

Figure 8-3   Energy levels for an organic dye.

S_0, S_1, and S_2 and the triplet states labeled T_1 and T_2.[2] Each band of energies consists of widely spaced vibrational levels and overlapping rotational levels. Radiative transitions are quantum mechanically forbidden between S and T levels, but transitions between one S level and another or one T level and another are allowed. The dye molecule is optically pumped to the S_1 level and drops to the lowest level in that band. Lasing takes place between that level and one of the many possible S_0 states. Continuous operation is not practical in most systems because intersystem crossing from S_1 to T_1 states, caused by collisions, results in a buildup of dye molecules in the T_1 states and optical absorption from T_1 to T_2 states. Pulse lengths are kept short compared with the intersystem crossing time of about 100 ns.

The bandwidth of dye lasers is 10^{13} Hz, which makes it possible to achieve picosecond pulses by mode locking.

8-5 RUBY LASER

The ruby laser was the first optical maser (or laser, as it came to be called) built. The first such laser was constructed by T. H. Maiman at the Hughes Laboratories, Malibu, California, in 1960. Figure 8-4 contains a schematic representation of a ruby laser.

The active material in the ruby laser is triply ionized chromium, Cr^{3+}, and the host is crystalline aluminum oxide (sapphire), Al_2O_3. Cr_2O_3 substitutes for about 0.05% by weight of the Al_2O_3, providing the crystal with a characteristic light pink color. The deep red color of gemstone ruby is the result of much higher concentrations of Cr_2O_3. The output of the ruby laser is at 0.6943 µm in the red.

Ruby lasers are operated in pulsed fashion with repetition rates on the

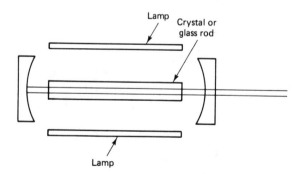

Figure 8-4   Schematic of solid laser, for example, ruby, Nd–Yag, Nd–Glass.

---

[2] Singlet and triplet states have zero and one, respectively, for the total spin of electrons in the molecule and determine the multiplicity of the energy levels.

order of one pulse per second. The energy per pulse may be as high as 100 J, but is usually a few joules. The normal pulse width is several milliseconds long, but Q-switching can be used to produce pulses of tens of nanoseconds to microseconds and correspondingly higher peak powers.

The ruby laser is a three-level laser and as such requires a high threshold for lasing action to occur, since over 50% of the Cr^{3+} must be raised to the excited state to achieve population inversion. Several joules of energy per cubic centimeter are required to establish a threshold-level population inversion. A great deal of the energy from the Xe flash lamps used to excite the laser goes into heating the laser rod; consequently, the duty cycle is limited. This problem is aggravated by the fact that ruby has a very low thermal conductivity. Cooling with refrigeration is used, but continuous operation is not practical.

The output of ruby lasers may be randomly or linearly polarized, depending on how the crystal is cut, that is, whether the c axis is parallel or perpendicular to the laser axis. The c axis parallel to the laser axis produces random polarization. When the c axis is perpendicular to the laser axis, linear polarization occurs.

The pulsed output of a ruby laser consists of random spiking due to heavily multimode operation. Q-switching is used to produce tens of megawatts peak power in 10- to 20-ns pulses. Ruby lasers are easily mode locked because of the multimode operation, and gigawatts of peak power in 10-ps pulses can be achieved.

The ruby laser was the first to be used for piercing holes in diamonds for wire-pulling dies. It can be used for piercing small holes in many materials and for spot welding. It is also used in applications where short, Q-switched pulses of light in the visible are required. One such application is in pulsed holography. By using high-energy submicrosecond pulses from a ruby laser, holograms of moving objects can be made, thereby eliminating the need for careful prevention of even the slightest motion during exposure, as required with low-power CW lasers.

8-6 NEODYMIUM–YAG LASERS

The neodymium (knee-oh-dim-e-umm)–YAG laser uses triply ionized Nd as the lasant and the crystal YAG (yttrium–aluminum–garnet) as the host. YAG is a complicated oxide with the chemical composition $Y_3Al_5O_{12}$. The amount of metal ions replaced by Nd^{3+} is 1% to 2%. The output wavelength is 1.06 μm, which is in the near ir.

Nd–YAG lasers are capable of average power outputs up to 1000 W. YAG lasers are operated continuously up to a few hundred watts, but pulsed operation, at high repetition rates, is used at the higher power levels.

The YAG laser is a four-level system, which results in a lower threshold

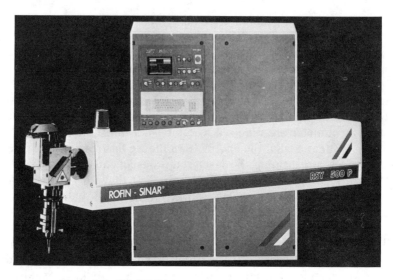

Figure 8-5   Nd–YAG laser. (Courtesy of Rofin-Sinar, Inc.)

energy to achieve the necessary population inversion. Also, YAG has a relatively high thermal conductivity; so, with cooling, CW or high-repetition pulsed operation is practical. CW YAG lasers are Q-switched with acousto-optical modulators to provide pulse rates of up to several thousand pulses per second, although average power begins to drop above 2000 pps.

Schematically, the YAG laser head appears similar to that of the ruby laser. Figure 8-5 is a photograph of a commercial Nd–YAG laser.

YAG lasers can be used for overlapping pulse seam welding, spot welding, hole piercing (including diamonds and other gemstones), and cutting. Cutting is accomplished by rapid overlapping hole piercing with a gas jet assist. Through frequency doubling and higher harmonic generation, YAG lasers are used for material property, photochemical, and interaction studies in the visible and ultraviolet ranges.

8-7 Nd–GLASS LASER

The Nd–glass laser uses Nd^{3+} as the lasant, but the host material is glass. The output is still 1.06 μm, although the bandwidth is several times that of the Nd–YAG laser, which partially explains a much higher pulse energy output. The large bandwidth results in hundreds of axial modes operating simultaneously. Mode locking puts all these modes in step, timewise and spatially, thereby resulting in a large bandwidth of about 10^{12} Hz and a correspondingly narrow pulse width of as low as 10^{-12} s.

High energy is achieved by using a physically larger active medium than in YAG lasers, which is practical with glass, for it can be made in excellent quality to almost any size. YAG crystals, on the other hand, are very expensive and difficult to grow in large size. Unfortunately, glass has a low thermal conductivity; as a result, Nd–glass lasers are limited to low-duty-cycle operation of about one pulse per second. The energy per pulse can be 100 J or more, making the Nd–glass laser suitable for many pulse-welding and hole-piercing applications.

One phenomenon occurring in solid lasers that can cause excessive beam divergence and self-focusing results from the fact that in some solids the refractive index increases with the irradiance of the beam. This creates a situation similar to that of a graded index optical fiber. The beam is actually guided by the self-induced refractive index change. If the irradiance exceeds a threshold level, the beam is focused to a small spot, usually damaging or destroying the rod.

8-8 EXCIMER LASERS

Although excimer stands for excited dimer, the molecules are not true dimers, but are in fact the complexes XeF, XeCl, ArF, and KrF. Hence, the correct term is excited complex, or exciplex for short. The term excimer is accepted, however.

Excimer lasers are unique in several respects. They are inherently pulsed because of the short lifetime of the excited state. Pulse lengths are typically tens of nanoseconds.

Despite the short lifetime of the lasing state, excimer lasers are very high gain systems. The devices typically have a rectangular-shaped transverse discharge region, similar to CO_2 TEA lasers. In fact, with some modification, excimer lasers can be used as either TEA or nitrogen lasers. Feedback is minimal, around 50%. Aluminized broadband reflectors can be used for the rear mirror. The optical configuration is generally two flat mirrors or an unstable resonator. The output beam is rectangular with about 3- to 5-mrad divergence. The unstable resonator configuration provides somewhat reduced divergence.

The active gas mixture is diluted by inert helium. Switching the laser from one gas to another is fairly simple if the same halogen is used. If a switch is made from F to Cl or Cl to F, then several hours of passivation are required to remove the halogen that was previously used, which is adsorbed to surfaces in the laser. It is easier to switch from Cl to F than from F to Cl because of the higher reactivity of F.

Because of the short pulses, low feedback, and highly multimode nature of excimer lasers, the output has low coherence, both spatially and temporally. The low spatial coherence has been used advantageously in the semiconductor industry. High spatial coherence causes excessive diffraction and associated loss of detail in certain photolithographic applications.

The short wavelength (high energy per photon) and extremely short pulse length of excimer lasers result in a very different type of interaction mechanism with matter than for visible or ir lasers. The photon energy at excimer wavelengths is sufficient to cause direct bond disruption in organic and some inorganic materials. Matter can be explosively removed with minimal thermal damage to surrounding material. This has led to applications to polymers, eye surgery, and the semiconductor industry. Excimer lasers are also used extensively in photochemical research and to pump dye lasers.

8-9 DIODE LASERS

Diode lasers, as might be guessed, are diodes that lase. There are about 20 known compound semiconductor diodes that lase with wavelengths ranging from 0.33 μm in the uv to nearly 40 μm in the ir. Electron beam excitation is used for all wavelengths shorter than 0.8 μm. In most of the others, e-beam or injection current (current in a forward-biased diode) can be used. For practical purposes, only the injection current type is discussed. The most common types of commercially available diode lasers are based on GaAs and ternary compounds, such as $Al_xGa_{1-x}As$. The $Al_xGa_{1-x}As$ compounds are discussed here as an example of diode lasers.

The first diode lasers operated in the early 1960s were simple homojunction GaAs devices, meaning that there was only one junction between the p- and n-type GaAs. Figure 8-6 is a schematic representation of a homojunction diode laser and the corresponding energy-level diagram. The cavity is formed by the cleaved ends of the GaAs chip, which is about 0.5 mm wide and 0.1 mm high. Radiation at about 0.9 μm is emitted from the depletion region. Contact is made through gold, Au, deposited on the anode and the Au substrate that the chip is mounted on. Notice that the Fermi level is in the bands in the energy-level diagram. This condition is indicative of degenerate (extremely heavy) doping approaching the solid solubility limit. Forward biasing of the diode results in a large injection current and a population inversion in the depletion region. The reason for the population inversion can be seen by inspection of Fig. 8-6. Only pulsed operation at liquid nitrogen temperature (77 K) is possible. Pulses with peak powers of a few watts are possible with repetition rates of several thousand pps.

Modern versions of the GaAs diode laser use single or double heterojunctions with stripe geometry and possibly large optical cavity (LOC) configurations to reduce the current density and risk of damage due to large radiation fields in the chip. A heterojunction refers to a junction between GaAs and $Al_xGa_{1-x}As$. A double heterojunction device with stripe geometry is depicted in Fig. 8-7. The purpose of the stripe contact is to improve the

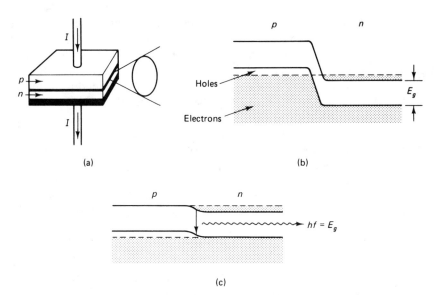

(a)   (b)

(c)

Figure 8-6 Homojunction diode laser: (a) diode, (b) energy bands, unbiased, and (c) energy bands, forward biased.

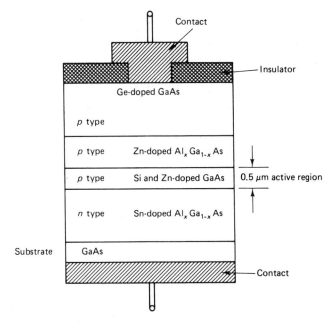

Figure 8-7 Double heterostructure diode laser with stripe geometry.

conduction of heat from the active region. The ternary compounds of $Al_xGa_{1-x}As$ have much poorer thermal conductivity than GaAs. Limiting the active region to a narrow stripe allows lateral heat conduction in GaAs to greatly reduce the temperature rise in the active region.

The purpose of the heterojunctions can best be explained in terms of the energy-level diagram shown in Fig. 8-8. Note that the energy gaps are wider in $Al_xGa_{1-x}As$ than in GaAs. This factor has two effects. First, the resulting energy barriers reflect both electrons and holes back into the active GaAs region, thus reducing the current density required to produce stimulated emission. Second, there is a refractive index difference between $Al_xGa_{1-x}As$ and GaAs (3.4 for $x = 0.4$ and 3.6 for $x = 0$), which causes light traveling at slight angles to the cavity axis to be strongly reflected back into the active region so that less radiation need be generated, thereby lowering the required current density. Devices of this sort can be operated CW at room temperature with the aid of a thermoelectric cooler to prevent an excessive temperature rise. The current density required is 1000 A/cm^2. If the bottom heterojunction is replaced by an ordinary pn GaAs junction, the result is a single-heterojunction device that can be operated in pulsed fashion at room temperature. Power levels are 10 to 20 W peak with repetition rates of up to 10^4 pps.

LOC devices are made with two additional layers $Al_yGa_{1-y}As$ on either side of the p-GaAs with $y \ll x$ so that the optical cavity becomes much wider, tens of micrometers compared to perhaps 1 μm, thus reducing the danger of damage to the crystal from the radiation. This feature also greatly reduces the diffraction of the beam as it leaves the end of the crystal from an angle of as high as 30° to one of only about 2°. Diode lasers achieve efficiencies of 4% to 5% for room temperature operation.

Variation of the amount of Al in these lasers permits adjustment of the emission wavelength from 0.84 to 0.95 μm.

The largest application of diode lasers appears to be in the field of communications, particularly fiber-optics communication. Because of the

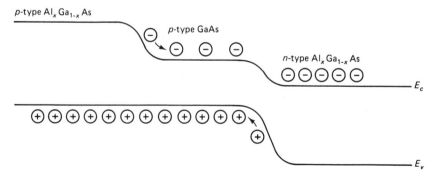

Figure 8-8 Energy-level diagram for double heterostructure diode laser.

great speed with which diode lasers can be switched on and off and their narrow bandwidth, they are excellent candidates for high-data-rate fiber-optic communications. Diode lasers will also undoubtedly find applications in pattern recognition, bar code reading, consumer products such as video disks, computer memory, and some ir illumination applications where fairly high power monochromatic radiation is needed.

8-10 CO₂ LASERS

The carbon dioxide (CO_2) laser is a molecular laser in which molecular vibrations rather than electronic transitions provide the mechanism for lasing action. Other lasers in this category are the carbon monoxide and hydrogen fluoride lasers. The carbon monoxide laser has achieved high efficiency, around 40%, but the corrosiveness and toxicity of carbon monoxide have prevented scaling of carbon monoxide lasers to the power levels achieved in CO_2 lasers. Hydrogen fluoride lasers have been primarily of interest in gas-dynamic and chemical laser work aimed at extremely high power levels (megawatts) for periods of time on the order of a second.

Carbon dioxide lasers have achieved success in a wide variety of industrial-material-processing applications at power levels from a few watts to 15 kW with reasonably high efficiency of 10% for commercial models.[3] The types of materials that the CO_2 laser is applied to include paper, wood, glass, plastic, ceramics, and many metals. Industrial processes include heat treating, welding, cutting, hole piercing, scribing, and marking.

Because of its wide industrial use, the CO_2 laser is discussed in some detail. Some of the CO_2 laser designs that are commercially available are qualitatively described. Some specific CO_2 laser applications, along with material-processing applications with other types of lasers, are presented in Chapter 13.

General description.   Here we consider the interactions that take place in the CO_2 laser. Pure CO_2 can be made to lase, but only weakly. Early in the development of CO_2 lasers, Kumar Patel discovered that a mixture of CO_2, N_2, and He greatly increased the power output. A typical ratio of CO_2:N_2:He partial pressures is 0.8:4.5:10. This will vary somewhat from system to system. The use of 10% oxygen mixed with the N_2 improves electrode and optics life without noticeably reducing power output.

The CO_2 molecule is a linear triatomic molecule and, as such, has only three distinct normal modes of vibration. These modes are shown schematically in Fig. 8-9. They are the asymmetrical stretching modes at $f_a =$

---

[3] This is overall efficiency defined as the ratio of the useful power output to the *total* power input, not just discharge power.

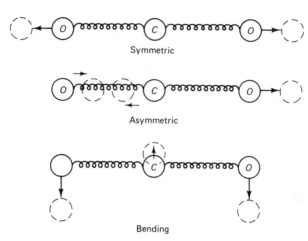

Figure 8-9 Vibrational modes of the CO_2 molecule.

7.0×10^{13} Hz, the symmetrical stretching mode at $f_s = 4.2 \times 10^{13}$ Hz, and the bending mode at $f_b = 2.0 \times 10^{13}$ Hz. The bending mode is degenerate in that it can occur in either of two perpendicular planes.

Molecular vibrations, like electronic states, are quantized, meaning that energy can be added to or removed from these vibration states only in discrete quanta of energy. The amount by which the energy can change in any given vibration mode is given by $\Delta E = \pm hf$, where f is the frequency of the particular vibration mode. This leads to an energy-level scheme for molecular vibrations similar to electronic energy-level diagrams. A simplified version of the CO_2 energy-level diagram is presented in Fig. 8-10, along with the first excited state of N_2. N_2 is a diatomic molecule; therefore, it has only one normal mode of vibration. The abbreviated spectroscopic notation in Fig. 8-10 for CO_2 (l, m, n) refers to the number of quanta in the various vibration modes, l quanta in the symmetrical mode, m in the bending mode, and n in the asymmetrical mode. Only those levels actively involved in the excitation and lasing process are shown in Fig. 8-10.

Note that N_2 has a near coincidence between its first vibration state and that of the (001) state of CO_2. This excited state of N_2 is extremely long lived, nearly ensuring that it will give up its energy, received by means of collisions with electrons, to an unexcited CO_2 molecule in a resonant or elastic energy transfer. Direct excitation of CO_2 molecules by collision with electrons is also important. Lasing can occur between the (001) and (100) states at 10.6 μm or the (001) and (020) states at 9.6 μm. The probability for the latter transition is only about one-twentieth that of the 10.6-μm transition, and so it does not occur normally unless the 10.6-μm transition is suppressed. From an industrial applications standpoint there is no advantage to operating at 9.6 μm, which would result in greatly reduced power output.

The vibrational states of the CO_2 molecule are split into many closely

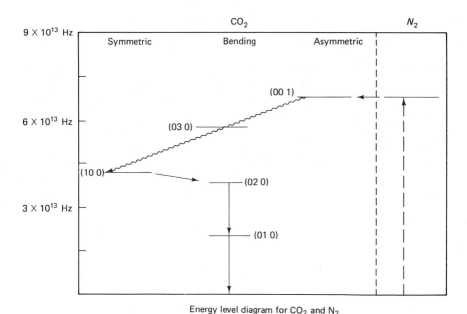

Energy level diagram for CO_2 and N_2

Figure 8-10   Energy-level diagram for CO_2 and N_2.

spaced energy states due to the quantized rotational states of the molecule. Quantum mechanical selection rules require that the rotational state either increase or decrease by only one quantum upon a transition from one vibrational state to another. That is, if J is the rotation quantum number, $\Delta J = \pm 1$. This situation leads to a large number of possible lines with two distinct branches, one for $\Delta J = 1$ centered at about 10.2 μm and one for $\Delta J = -1$ centered at 10.6 μm. Because of the slightly lower energy change for $\Delta J = -1$, this branch is favored, leading to a nominal output wavelength of 10.6 μm. Usually only one rotational transition operates because the one with highest gain is the most populated. Extremely rapid transitions between rotational levels occur due to thermal effects. The bandwidth of CO_2 lasers is about 60 MHz as a result of collisions and the Doppler effect.

The natural lifetime of the (001) state of CO_2 is on the order of seconds, but collision effects in the gas reduce this lifetime to the order of milliseconds, which is suitable for laser action. The collision lifetimes of the lower states are on the order of microseconds, but thermal effects repopulate the (010) state, creating a bottleneck in the return of CO_2 molecules to the ground state. Collisions between CO_2 molecules in the (100) state and unexcited molecules can cause transfer of energy to the (020) state. Subsequent collisions with unexcited CO_2 molecules can cause transfer to the ground state. Collisions with He atoms (which act as an internal heat sink), with the walls of a discharge tube, or with a heat exchanger, also can return excited molecules to the ground state. All CO_2 lasers of more than a few watts CW power output use some form of flowing gas technique to ensure a fresh supply of gas in the active region at all times. Many designs producing 1 kW or less simply exhaust the gas into the atmosphere. The total flow rate is around 0.1 SCMH.[4] The gas can be recirculated through a heat exchanger and is generally for higher-power-level machines. A small amount of gas is still continuously or periodically removed and made up by fresh gas to keep the gas mixture free of decomposition products, particularly carbon monoxide, which absorbs strongly at 10.6 μm and will quench lasing action. Such by-products are also harmful to the optics and electrodes.

In the following discussion of CO_2 lasers, low power refers to up to 200-W average power and high power refers to anything over 200 W.

Low-power CO_2 lasers. There are several types of CO_2 lasers that operate in this regime. High-power CO_2 lasers employ a flowing mixture of He, N_2, and CO_2. There are, however, sealed-tube designs that operate off 110 V that provide up to 100 W of power. After a few thousand hours of operating time, the tube must be recharged with the gas mix.

One version of the sealed-tube type is called a waveguide laser because a very narrow ceramic tube is used to contain the gas mixture, and the tube literally acts as a waveguide for laser radiation. The output for sealed-tube CO_2 lasers is of high quality and is very near pure TEM_{00}.

The term TEA laser stands for transverse-excited-atmospheric CO_2 laser. In these lasers the gas mixture is near atmospheric pressure in a rectangular-shaped transverse discharge. Operation is pulsed, with pulse lengths from nanoseconds to microseconds. Short pulse length is inherent in the transverse-discharge, high-pressure design of these lasers.

Low-power CO_2 lasers have many applications in industry, science, and medicine. Some examples are marking, soldering, paper etching and cutting, various surgical uses, and photochemistry.

High-power CO_2 lasers. Three basic types of CO_2 lasers are produced today for operation at 500-W CW or more. The differences in these types are significant enough that each is briefly described.

1. *Axial discharge with slow axial flow.* This type of laser is schematically represented in Fig. 8-11. The power capability of this type is 50 to 70 W/m, although efforts are being made to increase this amount through better heat transfer from the gas to the cooling liquid that surrounds the laser tube in a separate tube or jacket. Power outputs up to 1.5 kW are achieved by placing tubes in series optically, with electrical discharge and gas flow in parallel. That is, the electrical discharge and

---

[4] SCMH stands for standard cubic meters per hour.

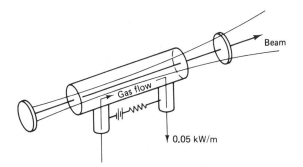

Axial discharge with slow axial gas flow

Figure 8-11 CO_2 laser with axial discharge and slow axial flow.

gas flow are independent for each tube that is 2 or 3 m in length. The tubes are usually mounted on a granite slab or in an Invar rod frame for mechanical and thermal stability. Figure 8-12 is a photograph of a 1-kW laser of this type.

Some of these lasers can be electronically pulsed to achieve much

Figure 8-12   Slow flow, 1.5-kW CO_2 laser capable of up to six times average power in pulse mode. (Courtesy of Coherent General.)

higher peak power without decreasing the average power significantly. This factor is advantageous in a number of material-removal operations. The tuber diameter of such lasers is usually quite small, resulting in the TEM_{00} mode being strongly favored. Somewhat larger bore designs are available if low-order multimode operation is desired, such as for heat treating.

2. *Axial discharge with fast axial flow.* The chief difference between this design and the previous one is that the gas mixture is blown through the tubes at high speed and recycled through a heat exchanger. Each CO_2 molecule is in the tube only long enough to lase and then passes through the heat exchanger. The power output of this type of laser is 600 W/m. The tubes may be of fairly large bore; consequently, either unstable or higher-order-mode outputs are common. Gaussian operation is reported for several kilowatts. Such lasers are available from 500-W to 6-kW CW power output with the potential for much higher power levels. Figure 8-13 is a photograph of an axial discharge, axial fast flow laser.

3. *Transverse discharge, transverse fast flow.* This type of laser uses a fast transverse gas flow, about 60 m/s, with recirculation, like the previous design, but with a conventional electrical discharge transverse to the flow and the beam. The discharge takes place between a hollow water-cooled cathode or pins and a water-cooled segmented anode, with each segment of the anode individually ballasted by a resistor. Relatively

Figure 8-13   Fast axial flow laser (1.5 kW). (Courtesy of PRC Corp.)

Figure 8-14   3.0-kW transverse discharge, transverse fast flow CO_2 laser. (Courtesy of Spectra Physics.)

low voltage sustains the discharge at very high current because of the short distance between the cathode and the anode. A power output of about 600 W/m is achieved. The beam is folded back and forth through the discharge region five to seven times, resulting in an actual power output of about 2.5 kW for a laser with approximately 1.2 m separating the cavity mirrors. Figure 8-14 shows a photograph of one such laser.

The outputs of these lasers can vary from nearly Gaussian to very high order or may be unstable, depending on the optical configuration of the cavity. Lasers of this type are available with power outputs from 1 to 9 kW, and much higher power levels are attainable.

PROBLEMS

8-1. Discuss neutral gas, ion, and metal vapor lasers with particular emphasis on their differences.

8-2. Discuss dye lasers with particular emphasis on the pumping schemes and the reason for pulsed operation.

8-3. Discuss and compare ruby, Nd–YAG, and Nd–glass lasers.

8-4. Discuss the diode laser with comparisons between homojunction and heterojunction devices and the methods used to reduce injection current density.

8-5. Discuss CO_2 laser operation and describe and compare three types of CO_2 laser designs.

8-6. Construct a table of pertinent data, such as wavelength, power or pulse energy range, bandwidth, and type of applications, for six of the lasers discussed in this chapter.

8-7. Explain how excimer lasers differ from other gas lasers and discuss the unique differences between excimer laser beam properties and most other lasers.

REFERENCES

CHARSCHAN, S. S. (ed.), *Lasers in Industry*. New York: Van Nostrand Reinhold Co., 1972.

DEMARIA, A. J., "Review of CW High-Power CO_2 Lasers," *IEEE Proceedings*, **61**, No. 6, June 1973.

DULEY, W. W., *CO_2 Lasers: Effects and Applications*. New York: Academic Press, Inc., 1976.

HAUS, HERMANN A., *Waves and Fields in Optoelectronics*. Englewood Cliffs, NJ: Prentice Hall, 1984.

MOSS, T. S., G. J. BURRELL, AND B. ELLIS, *Semiconductor Opto-Electronics*. New York: John Wiley & Sons, Inc., 1973.

READY, J. F., *Industrial Applications of Lasers*. New York: Academic Press, Inc., 1978.

SILVAST, W. T., "Metal-vapor Lasers," *Scientific American*, February 1973.

SVELTO, O., *Principles of Lasers*. New York: Plenum Press, 1982.

VERDEYEN, J. T., *Laser Electronics*. Englewood Cliffs, NJ: Prentice Hall, 1981.

Chapter 9

Using Low-Power Lasers for Alignment, Gaging, and Inspection

Industrial systems that use low-power lasers for alignment, gaging, or inspection generally consist of four basic units. These units are a laser light source, an optical system to direct and structure light, an optical system to collect or image the light after it has interacted with the object or medium being interrogated, and a detection system. The detection system may be a human observer or an optoelectronic device interfaced with an electronic display unit or computer. This chapter describes the basic principles of low-power laser systems used in industry. Some comments on using a computer to interpret optical information are made, but basically this topic is beyond the scope of this book.

The applications of low-power laser systems discussed in this chapter do not use interferometry. Most systems are simple in nature and can be designed to operate in an industrial environment. More complex applications, such as holographic and speckle interferometry, discussed in Chapter 11, usually require the protection of a laboratory to operate in a reliable fashion.

9-1 ALIGNMENT

Perhaps the most obvious application of a low-power laser is alignment. In this application the laser beam is used simply as a pointer to align such things as drainage tile, ditches, building foundations, or industrial equipment. In some applications it is important to align an object so it has the proper position and angular alignment. An example is spindle alignment. Assume in a machining operation that a drill is mounted on a spindle to drill a hole in a part on a transfer line. Laser alignment systems are frequently used to align the spindle at the proper position and angle relative to a hole in a master part. Most alignment systems consist of a transmitter, receiver, and a display or computer system used to communicate with the operator.

Transmitter. Transmitters used for alignment with low-power lasers usually consist of the laser source and optics to structure the light. HeNe and diode lasers are commonly used. HeNe lasers are larger in physical size than diode lasers, but have the advantage of producing a visible beam that requires less optics to structure the light. Visible-light diode lasers are now available, but many first-generation alignment systems used diode lasers operating at infrared wavelengths.

A diode laser output window is extremely small and resembles a rectangular aperture that is longer than it is wide. Thus the output beam from a diode laser has a large divergence angle, and in the far field the cross section of the beam is elliptical. To structure the output beam for alignment applications, the laser beam is first collimated using a converging lens. This collimated beam still has an elliptical cross section. To circularize the beam cross section, a pair of cylindrical lenses or prisms can be used. As illustrated in Fig. 9-1, these devices act as a one-dimensional beam expander to stretch the beam in one direction so that the cross section of the beam is circular.

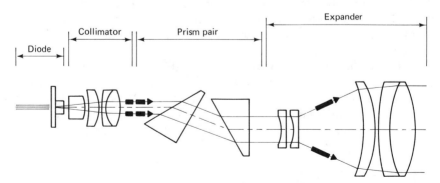

Figure 9-1 Optics used to circularize and collimate beam from diode laser. (Courtesy of Melle Griot.)

The output beam of a HeNe laser operating in the TEM_{00} mode has a circular cross section and does not require the aforementioned optics.

If alignment is required over a long distance, a circularized diode or HeNe laser beam can be expanded further to reduce beam divergence, as illustrated in Fig. 9-1. The other advantage of expanding the beam further is that it presents less potential hazard to the human eye. It is extremely important, even when working with low-power lasers, that current laser safety standards be followed and that lasers be used safely.

Detector.   In many alignment applications a visible laser beam from a transmitter is incident on the object to be aligned. A human observer then moves the object to be aligned so that the spot of light on the object due to the laser beam is in the proper location. By using a quadrant detector, objects can be aligned relative to the laser beam with greater accuracy. As discussed in Chapter 4 and illustrated in Fig. 9-2, the signal output from each quadrant is identical when the laser beam is centered on the detector. If the laser beam or object moves so that more power is incident on one or two diodes, there will be less power incident on other diodes. The signal outputs from the quadrants can be used to determine the position of the laser beam relative to the quadrant detector center. Note that the diameter of the laser beam must be large enough so that some light power is incident on all quadrants during the alignment process, but less than one-half the diameter of the detector.

In many applications, angular alignment is also important. To describe the process of measuring angular alignment, we will first assume that a laser

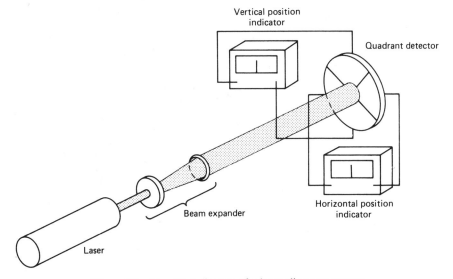

Figure 9-2 Simplified diagram of a laser alignment system.

beam incident at an angle θ relative to a quadrant detector center line is centered on a quadrant detector. A converging lens is then placed in front of the detector, with its optical axis along the center line of the detector and one focal length F away from the detector, as illustrated in Fig. 9-3. The laser beam will be focused to a point a distance

$$y = F\theta \qquad (9\text{-}1)$$

from the center of the quadrant detector. This focused light cannot be used by the quadrant detector to measure θ because light is no longer incident on all quadrants. To increase the diameter of the spot of laser light on the detector, a diverging lens of focal length $-F/2$ can be placed on the optical axis of the converging lens and a distance $F/2$ from the detector. The diameter of the spot of light on the detector will be one-half the laser beam diameter without the two lenses, but has sufficient size to illuminate all quadrants of the detector so that a measurement can be made.

Modern laser alignment systems can use the output signals from a quadrant detector and determine the distance at which laser light strikes the

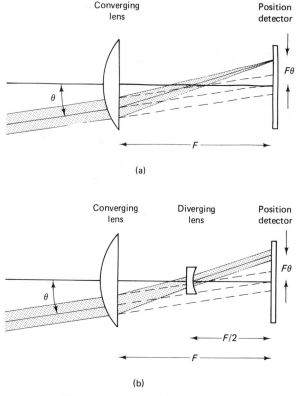

(a)

(b)

Figure 9-3 Angular alignment optics.

detector relative to its center in millimeters or inches. By using the proper optics in front of the detector and measuring the distance at which the laser beam strikes the detector relative to its center, a computer can use this distance and display an angular alignment in a variety of units.

9-2 SCANNER SYSTEMS

Optical devices used to scan laser beams include mirrors, prisms, and holograms. Mirrors and prisms are generally oscillated sinusoidally or rotated with a constant angular velocity. Galvanometer mirror scanners can respond in a linear manner to input current. By using two of these devices, a laser beam can scan a two-dimensional surface. After this scanned light interacts with an object, it can be used in applications such as leveling, gaging, surface inspection, and bar code reading.

Leveling.   Figure 9-4 is a simplified diagram of a laser leveling device. A beam expander is first used to increase the diameter of the HeNe laser beam. A penta-prism is used to turn the laser through a right angle. A penta-prism turns the beam through a right angle even though the prism orientation is not precisely controlled. Rotating the prism about the axis of the incident beam rotates the beam so it sweeps out a flat plane.

Scanner systems of this type are routinely used in the installation of walls and ceilings. By leveling the scanning beam at ceiling height, the beam can be used to align the framework for ceiling tile. The alignment is accomplished by the construction worker viewing the reflected laser light from a target or the framework being installed. By scanning out a vertical plane,

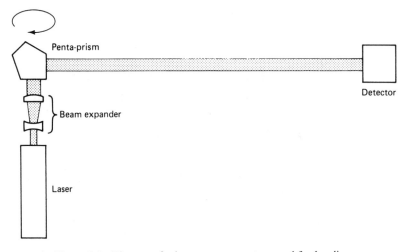

Figure 9-4   Diagram of a laser scanner system used for leveling.

the scanning beam can be used to align wall studs. Detectors can also be used to align objects relative to the scan plane. A common application is leveling forms used to contain concrete when it is poured in its liquid state before it hardens. In this application a detector is mounted on a stake held vertically, with its base placed on top of the concrete form. A detector is used near the top of the stake to detect the plane scanned out by the laser beam. By maintaining a constant distance between the scan plane and the top of the form, it can be leveled. To grade a level surface with earth-moving equipment, systems have been designed to control the blade of the earth-moving equipment to remove the proper amount of dirt so that the graded surface is flat. These systems are used primarily in road construction and agriculture.

Gaging. Figure 9-5 is a simplified diagram of a laser scanner system that can be used for gaging. This system consists of a transmitter and receiver. In the transmitter unit a laser beam from a HeNe laser is incident on a mirror rotating at a constant angular velocity ω_M. The reflected beam is scanned across a diameter of a converging lens. By placing the rotating mirror at the front focal point of the lens, the beam will emerge from the lens and thus the transmitter parallel to the lens axis. The object used in our example is an opaque cylinder placed in the scan plane of this beam; the laser beam will be blocked for a period of time Δt as the beam scans across

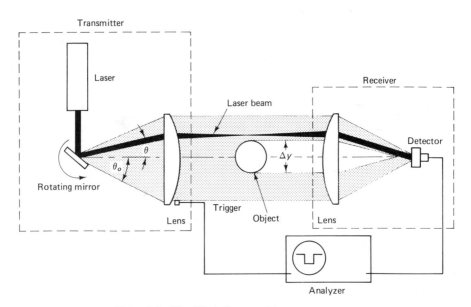

Figure 9-5 Simplified diagram of laser scanner gage.

it. The receiver contains a second converging lens, which is used to focus this beam onto a photodiode. By timing the time period Δt that the beam is blocked, the cylinder diameter Δy can be determined.

In the transmitter unit, the laser beam reflected from the rotating mirror located at the front focal point of converging is reflected at an angle θ with respect to the lens axis. If this lens is designed to meet the Abbe sine condition, the beam will emerge with a displacement

$$y = F \sin \theta \tag{9-2}$$

where F is the lens focal length. Angular displacement θ, in terms of mirror angular velocity ω_M, is

$$\theta = 2\omega_M t - \theta_0 \tag{9-3}$$

The initial angle $-\theta_0$ is the angular displacement where the laser beam scans a trigger photodiode and time is assumed to be zero. Note that the angular velocity of the laser beam ω_B is two times the angular velocity of the mirror ω_M. The diameter of the cylinder Δy, illustrated in Fig. 9-5, can be found by using

$$\Delta y = F[\sin (2\omega_M t_2 - \theta_0) - \sin(2\omega_M t_1 - \theta_0)] \tag{9-4}$$

where t_1 and t_2 are the times when the laser beam scans past the edges of the cylinder. The transverse velocity v of the beam emerging from the lens is

$$v = \frac{dy}{dt} = 2\omega_M F \cos(2\omega_M t - \theta_0) \tag{9-5}$$

For small angular displacements, Eq. (9-2) becomes

$$y = F\theta \tag{9-6}$$

Eq. (9-4) reduces to

$$\Delta y = 2\omega_M F \Delta t \tag{9-7}$$

And Eq. (9-5) to

$$v = 2\omega_M F \tag{9-8}$$

Thus, by using small angular displacements and rotating the scanner mirror with a constant angular velocity, the transverse velocity of the laser beam emerging from the transmitter is considered to be constant.

Upon emerging from the transmitter, the laser beam is not only parallel to the lens axis but also converges to a waist. By centering the object to be gaged along the line scanned by the laser beam waist, the time required for the beam to scan past an object edge is reduced to a minimum. Assuming that the output from the receiver is directly proportional to the laser beam power incident on it, this output is as illustrated in Fig. 9-5. The laser beam is blocked by the object for a time interval Δt. It is interesting to note that if the object edge can be considered to be a sharp edge, the time interval should be measured between the points when the laser beam power incident on the photodiode is approximately one-quarter of its maximum value [Goodman, 1968].

Again, it is important to realize that this gaging technique is only accurate as long as mirror angular velocity ω_M is constant and the lens meets the Abbe sine condition. For small angular displacements θ, Eq. (9-4) can be reduced to Eq. (9-7).

To remove a number of the restrictions that we have imposed on the previous system, parabolic mirrors can be used rather than lenses. Figure 9-6 is a simplified diagram of such a system. The laser beam first passes through a converging lens to structure the light so a beam waist is located at the object position. Collimated light incident parallel to the optical axis of a parabolic mirror is focused without spherical aberration. Thus larger scan planes can be obtained. In the illustration in Fig. 9-6, the mirror scanner is placed at the focal point of the parabolic mirror. The light reflected from the parabolic mirror has the displacement of

$$y = 2F \tan \left(\frac{\theta}{2} \right) \tag{9-9}$$

where F is the focal length of the parabolic mirror (see Problem 9-7). The diameter of a cylinder placed in the scan plane can be determined by using

$$\Delta y = 2F \left[\tan \left(\omega_M t_2 - \frac{\theta_0}{2} \right) - \tan \left(\omega_M t_1 - \frac{\theta_0}{2} \right) \right] \tag{9-10}$$

where t_1 and t_2 are the times that the laser beam scans the edges of the cylinder. The scan velocity is

$$v = 2\omega_M F \sec^2 \left(\omega_M t - \frac{\theta_0}{2} \right) \tag{9-11}$$

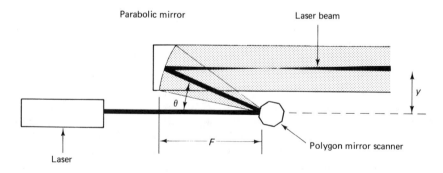

Figure 9-6   Scanner system with a parabolic mirror.

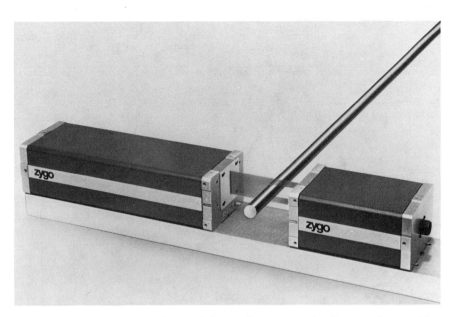

Figure 9-7   Laser scanner and detector being used to measure the diameter of a round bar. (Courtesy of Zygo.)

As in the case of the previous system, the equations for this system reduce to Eq. (9-6), Eq. (9-7) and Eq. (9-8) for small angular displacements. The n-sided polygon mirror, illustrated in Fig. 9-6, scans the laser beam n times across the parabolic mirror per revolution. An off-axis parabolic mirror can be used to direct and focus the laser beam without it being blocked by the scan mirror. Systems of this type can be used to produce a large scan plane because the parabolic mirrors eliminate spherical aberration. Figure 9-7 is a photograph of a laser scanner system being used to measure the diameter of a rod. To eliminate random fluctuations in the measuring process, commercial systems use a computer interfaced with the gage to average the measurements made by a number of scans that can be specified by the user.

Surface inspection.   Surface inspection with low-power lasers is done by evaluating the specular or diffuse reflection from the surface being interrogated. Laser scanner systems have been highly successful for inspecting machined surfaces for defects if the surface is specular and the defects are filled with black residue from the machining process. Figure 9-8 illustrates a scanner system that is designed to detect surface defects and the presence of a honed surface.[1] If the machined surface is highly specular in nature,

---

[1] *Honing* refers to the machining of a surface with hone stones or emery cloth, which produces very fine periodic scratches.

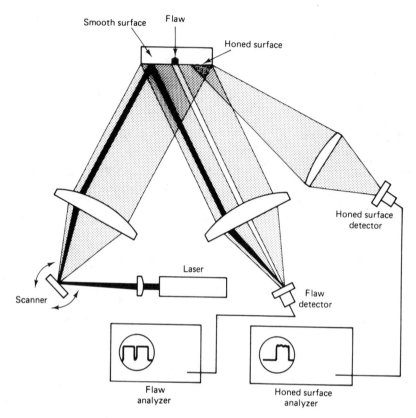

Figure 9-8   Laser scanner system designed to detect surface defects.

most of the laser light energy is reflected to the lens–photodiode detector positioned to receive the specular reflection. When the laser beam scanning the machined surface encounters a surface flaw, most of the energy is absorbed by the embedded residue in the flaw and specular reflected energy decreases. This drop in energy can be detected by an oscilloscope or electronics interfaced with the photodetector positioned to receive the specular reflected light. To obtain a significant decrease in specular reflected energy, the spot size of the laser beam on the surface should be smaller than the smallest dimension of the flaw being interrogated. The presence of hone will not significantly reduce specular reflected light energy but will increase the percentage of scattered light. By using a second detector system to detect this increase, the presence or absence of hone can be detected. Figure 9-9 is a diagram of a commercial scanner system designed to inspect for defects in sheets of material several meters wide. To inspect sheets with a specular surface finish for surface defects, the laser scanner and detector system would be mounted above the sheet stock as illustrated. Surface flaws such as rust, scratches, and pits will scatter some light toward the detector, whereas specular reflected light will not be received by the detector. The detector consists of a pair of cylindrical lenses to focus the light on the back surface of a glass or plastic tube. The back side of the tube is diffusive. Light is diffusely reflected at the back surface and then transmitted, because of internal reflection, down the length of the tube to a photomultiplier mounted on one end of the tube. A computer is used to process the output signals from the photomultiplier to indicate surface defects. Transparent sheets can be inspected by mounting the detector so that it receives the light transmitted through the sheet.

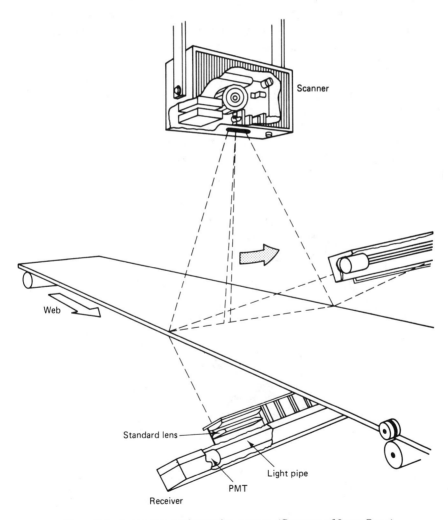

Figure 9-9 Laser scanner inspection system. (Courtesy of Intec Corp.)

Bar-code scanners.   Bar codes are being used extensively in industry for part identification. The code usually consists of black parallel bars of prescribed widths printed on a white background. A binary code can be represented by black bars of two different widths, a wide bar representing 1 and narrow bar 0. Many codes use both white and black bars of different widths to represent numbers and/or letters. Reading codes requires that the relative bar widths be measured relative to a set of reference bars at the beginning and end of the code.

To read the code, a laser beam is scanned across the black and white bars. As the beam scans across the bar code, much of the light is scattered by the white bars and most of the light is absorbed by the black bars. Since the relative widths of the bars are to be measured, the laser beam need not scan the code along a line perpendicular to the bars. The laser beam can scan across the bars at any angle as long as all bars are scanned. Most systems scan the beam in a crisscross pattern to ensure that the laser beam scans the entire bar code. Figure 9-10 is a sketch of a system that could be used as a bar-code reader. The laser beam is reflected from a mirror on a transparent window to two scanner mirrors and is then focused on a bar code by a converging lens. One scanner mirror is used to scan the beam in the vertical direction and the other in the horizontal direction. The converging lens is used to focus the beam to a waist at the bar-code location.

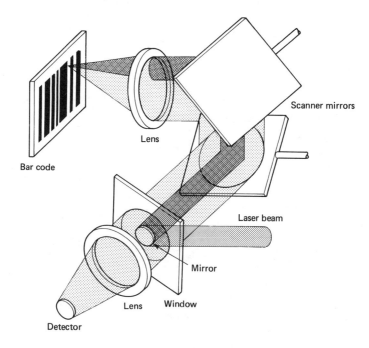

Figure 9-10 Bar-code scanner/reader.

Light scattered from the bar code is received back through the converging lens and reflected back to the transparent window by the scanner mirrors. Much of this scattered light is transmitted through the window around the perimeter of the mirror. This light is focused on a detector by a converging lens. The output signal from this detector can be used to interpret the bar code.

The science of bar-code design is extensive and will not be discussed in this text. It should be noted, however, that one of the most innovative applications of holography is to use a rotating holographic optical element (HOE) to scan a laser beam across a bar code. Holography will be discussed in Chapter 11.

9-3 OPTICAL TRIANGULATION

Optical triangulation provides a noncontact method of determining the displacement or profile of a diffuse surface. Two popular configurations use a spot of light or line of light projected on the diffusive surface.

Spot-of-light system.   Figure 9-11 is a simplified sketch of a spot-of-light triangulation system. A diode or HeNe laser and associated optics are used to produce a spot of light on a diffusive surface. A portion of the light is scattered from the surface and is imaged by a converging lens on a linear

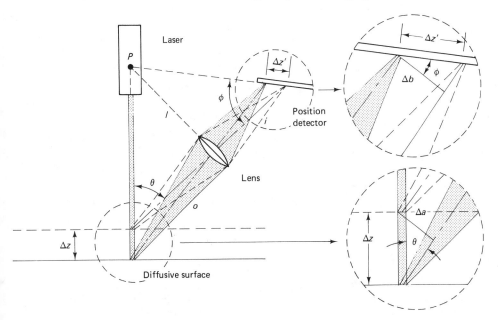

Figure 9-11 Diagram of optical triangulation system.

diode array or linear position detector. In our discussion, we will first assume that the laser beam is incident normal to a diffusive surface. If the diffusive surface is displaced parallel to the laser beam, the spot of light on the surface will have a component of displacement perpendicular to the optical axis of the converging lens. This displacement causes a corresponding displacement of the image on the detector. This image displacement can be used to determine the displacement of the diffusive surface. Note that, to maintain best focus of this image on the detector, the optical axis of the lens is not necessarily perpendicular to the detector.

To determine the proper angle between the optical axis and position detector, the Scheimpflug optical design condition can be used. The Scheimpflug condition states that object plane, the plane of the lens, and the image plane should all intersect at a line. For our application, this condition indicates that, to maintain best focus of the laser beam spot on the detector, the laser beam, lens plane, and the plane of the detector must all intersect at a common point P (Fig. 9-11).[2] In our design, we have placed the lens plane perpendicular to the optical axis. Under this condition, the angle between the laser beam and lens optical axis θ and the angle between the optical axis and position detector ϕ are related to the object distance o and image distance i by

$$l = i \tan \phi = o \tan \theta$$

where l is the distance between the point P and lens center. Thus angle ϕ can be determined by using

$$\tan \phi = \frac{1}{|m|} \tan \theta \qquad (9\text{-}12)$$

where the magnification $m = -i/o$. To facilitate our discussion of the equation used to determine the displacement of a diffusive surface, two expanded views have been provided in Fig. 9-11. In our derivation it is assumed that the displacement component of Δz parallel to the laser beam is small relative to the object distance o. Using this assumption, light rays shown in the expanded views, which represent light scattered toward the converging lens and light incident on the detector, are assumed to be parallel. Thus $\Delta a = \Delta z \sin \theta$ and $\Delta b = \Delta z' \sin \phi$. Transverse magnification amplitude due to the lens $|m| = \Delta b / \Delta a$. Using these equations, the image displacement on the detector $\Delta z'$ is found to be

$$\Delta z' = |m| \left(\frac{\sin \theta}{\sin \phi} \right) \Delta z \qquad (9\text{-}13)$$

---

[2] Note that the Scheimpflug principle indicates that the lens need not be perpendicular to the line connecting the laser light spot on the diffusive surface and the position detector center. We have assumed in our discussion that the lens is perpendicular to this line, which also serves as the lens axis.

Due to our approximations, Eqs. (9-12) and (9-13) are not exact equations. They are approximations that can be used to design a spot-of-light triangulation gage. Once built, the gage can be interfaced with a computer system and calibrated to obtain maximum accuracy. Figure 9-12 is a photograph of an optical triangulation system configured in the manner discussed. The results of Problem 9-9 provide a more general set of equations for optical triangulation gages:

$$\tan \phi = \frac{1}{|m|} \tan(\theta + \beta) \qquad (9\text{-}14)$$

and

$$\Delta z' = \left[\frac{|m| \sin(\theta + \beta)}{\sin \phi \cos \beta} \right] \Delta z \qquad (9\text{-}15)$$

To obtain this set of equations, it is assumed that Δz is parallel to the surface normal, β is the angle of incidence of the laser beam, and θ the angle between the lens optical axis and the surface normal.

Line-of-light system.   Figure 9-13 is a simplified sketch of a line-of-light optical triangulation system. To structure the light into a line on a diffusive surface being investigated, a cylindrical lens can be used. Some of the light scattered from this surface is imaged onto a two-dimensional diode array by a converging lens. The object chosen for our illustration is a plate

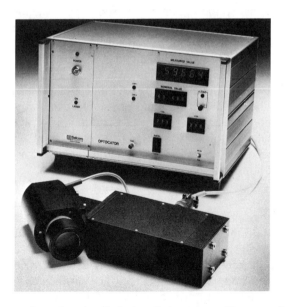

Figure 9-12   Laser-based optical triangulation system. (Courtesy of Selcom.)

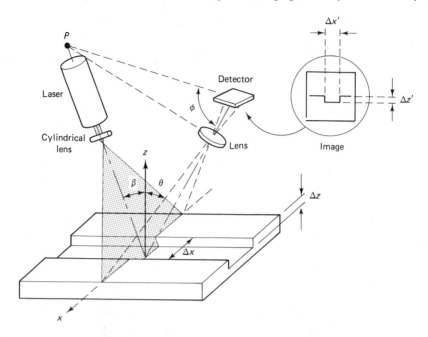

Figure 9-13   Line-of-sight optical triangulation unit.

with a rectangular groove with a width Δx and depth Δz. The image produced on the detector will have a discontinuity, as illustrated in the expanded view in Fig. 9-13. The short line segment represents the image of the line of light on the bottom of the groove, and the other two line segments represent the image of the structured light on the top surface of the object. The width of the short line segment $\Delta x'$ and the amount the short line is shifted from the other two line segments $\Delta z'$ are related to Δx and Δz, respectively. The relationship between groove depth Δz and image displacement $\Delta z'$ is given by Eq. (9-15). The angle ϕ that should be used to maintain best focus of the image on the detector is given by Eq. (9-14). The line of light and image are both perpendicular to the optical axis of the converging lens. Thus the relationship between $\Delta x'$ and Δx can be found by using the definition of transverse magnification. That is,

$$\Delta x' = |m| \Delta x \qquad (9\text{-}16)$$

Line-of-light optical triangulation systems require the use of a computer system to evaluate the image provided to the detector and then determine the actual dimensions of the object being investigated. One common application of these devices in the automotive industry is measuring fit-up between various car-body components. Surface profiles can be measured as well. By using two or more projected lines of light, systems of this type can be designed to make three-dimensional measurements.

9-4 OPTICAL RADAR

An alternative to optical triangulation is range finding by using optical radar. Optical radar is frequently referred to as lidar. *Radar* is an acronym formed from the words *ra*dio *d*etection *a*nd *r*anging. Radar can be defined as a device that consists of a transmitter that produces high-frequency radio waves and a detector that receives this radiation after being reflected from an object. By evaluating the round-trip time of the radiation, the object range can be determined. Object direction relative to a radar unit is determined from the direction of the transmitted and received radio waves. *Lidar* is an acronym formed from the words *l*ight *d*etection *a*nd *r*anging. Lidar systems use several different techniques to measure round-trip times. To measure large distances, pulsed lasers are frequently used. Retroreflectors were placed on the moon during one of the *Apollo* missions. By measuring the time of flight required for the light pulse to travel to the moon and back, the distance between the earth and moon can be measured with an accuracy of less than 1 meter. For industrial applications this technique is not sufficiently accurate. A more accurate technique is to measure the phase shift of an amplitude-modulated laser beam relative to a reference.

Figure 9-14 is a simplified drawing of an optical radar sensor system that determines range by using an amplitude-modulated laser beam. In our illustration the laser beam is produced by a diode laser driven by a sinusoidal

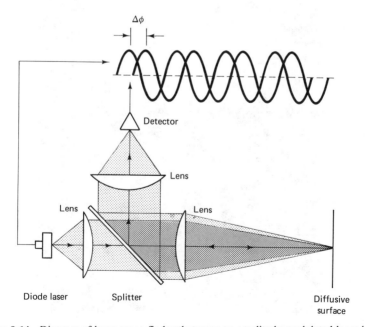

Figure 9-14 Diagram of laser range finder that uses an amplitude-modulated laser beam.

input current that produces a modulated laser beam. After being collimated, this beam passes through a splitter and is focused onto a target by a converging lens. Some of the light reflected from the target enters the converging lens, is reflected from the splitter, and is focused onto a photodiode. The input current to the photodiode and the output signal from the photodiode are compared, and the phase difference ϕ between these two signals is measured. It is assumed that a phase difference of zero corresponds to a range of zero; the phase shift ϕ in terms of the range R is given by

$$\phi = k_m(2R)$$

where $k_m = 2\pi/\lambda_m$ and λ_m is the modulation wavelength. The speed of light $c = \lambda_m f_m$, where f_m is the modulation frequency. Thus range can also be expressed as

$$R = \frac{c}{4\pi f_m} \phi$$

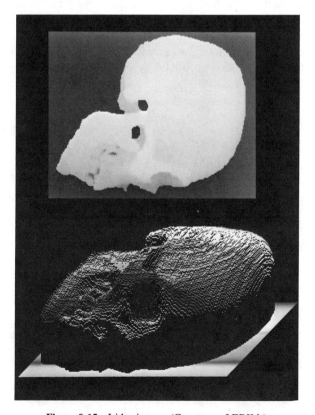

Figure 9-15   Lidar image. (Courtesy of ERIM.)

Single frequency modulation systems are limited to phase measurements between 0° and 360°. Thus range is determined by using

$$R = R_0 + \frac{c}{4\pi f_m} \Delta \phi \qquad (9\text{-}17)$$

where R_0 is an initial range where the signals are in phase. The ambiguity interval or the depth of field is $c/2f_m$.

The accuracy of the range measurement depends on the accuracy with which phase shift $\Delta \phi$ can be measured. Resolutions of a fraction of a millimeter have been achieved. By scanning the laser beam, three-dimensional measurements can be made. This technique shows excellent potential in the field of three-dimensional computer vision. Figure 9-15 is a photograph of an image obtained from such a system. This photograph is similar to that obtained by using a standard photographic camera except image brightness is a function of object range. Bright regions are closer to the lidar unit and dark regions farther away.

9-5 LINEAR DIODE ARRAY SYSTEMS

Many optical inspection devices found in industry use a camera lens to focus the image of an object on a one- or two-dimensional diode array. In this text we will concentrate on systems using a linear diode array (one-dimensional array). Two-dimensional diode array systems and their applications are covered in several good textbooks on computer vision. White-light illumination is usually preferred in these applications because laser speckle can produce a nonuniform image irradiance. Laser illumination is often used when the light must be highly structured or collimated. Unless otherwise stated, we will assume that a laser is being used as the light source. As in other applications, backlighting should be used whenever possible and reflected light from the object used when necessary.

A typical diode-array camera system using laser light to backlight an object is illustrated in Fig. 9-16. A beam spreader collimates and increases the diameter of the laser beam. A camera lens images the light passing above and below the object on the linear diode array. The output of the diode array can be displayed on an oscilloscope as illustrated in Fig. 9-16. Those diodes not illuminated produce little or no signal, and those diodes that are illuminated produce positive pulses that are displayed on the oscilloscope as the diodes are scanned in sequence. By counting the number of diodes not illuminated and taking into account the magnification due to the lens, the object width can be determined. For this system to be used for gaging, the lens must be corrected for aberrations, particularly field curvature and distortion. That is, the lens must be designed so that a linear relationship exists between measurements made in the object and image planes. In this text,

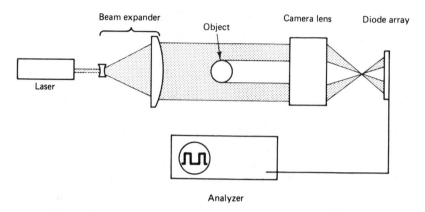

Figure 9-16   Typical linear photodiode array camera system.

the resolution with which a measurement of the image on the linear diode array can be made is assumed to be plus or minus the distance between two adjacent photodiodes. Thus the theoretical resolution with which a measurement of the object can be made will be considered to be the distance between two photodiodes divided by the magnification.

Using collimated light to backlight the object has an advantage over backlighting the object with diffused light. To demonstrate this concept, it is convenient to use the Newtonian form of the lens equations. Recall that this form of the lens equation states that

$$zz' = F^2 \tag{9-18}$$

where z is the distance between the object and the primary focal point, z' is the distance between the image and the secondary focal point, and F is the focal length of the lens as illustrated in Fig. 9-17.

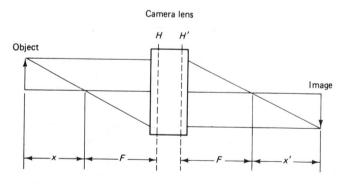

Figure 9-17   Notation for Newtonian form of lens equations.

Magnification is

$$m = \frac{y'}{y} = -\frac{z'}{F} = -\frac{F}{z} \tag{9-19}$$

where y and y' are the object and image heights, respectively. If collimated laser light is used to backlight the object, $-z'/F$ can be used to determine magnification. If the object is moved from the ideal object distance $z + F$ by a distance Δz, the magnification will remain the same as long as z' is held constant. This concept is illustrated in Fig. 9-18.

If the image of the object edges are observed, the slope of the image irradiance versus position (y) curve will have its largest value. If the object is displaced from this ideal distance z by an amount Δz, the image will not be as well focused. That is, at the image of the object edges the slope of the image irradiance versus position (y) curve will be smaller, but the magnification will not change. If white light is used to backlight the object, the locations of object edges correspond to the inflection points of the irradiance versus position (y) curve of the image. This inflection point is also considered to be the point at which the irradiance is the average of the maximum and minimum values of the irradiance on either side of the edge image. If coherent light is used, diffraction accrues, and the apparent position of the image of

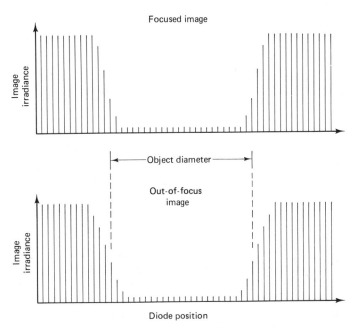

Figure 9-18 Focused and out-of-focus digitized images produced using a linear photodiode array, collimated light, and a constant image distance.

the edges changes and is located at a point where the irradiance is less than the average of the maximum and minimum values of the image irradiance. If the average value is used, slight errors can accrue [Goodman, 1968].

Two-dimensional photodiode arrays are now being used as detectors in computer vision systems. In these systems, image information is digitized and presented to a computer to be processed and evaluated. Applications of computer vision in industry include inspection, robot guidance, and process control. The topic of computer vision is too extensive to be covered in this text. A reference is listed at the end of this chapter.

PROBLEMS

9-1. Assume that the quadrant detector illustrated in Fig. 9-2 has a diameter of 1.0 cm and the laser beam incident on the detector has a diameter of 0.5 cm.
(a) Determine the maximum displacement that can be measured.
(b) Assume that F for the confocal lens system illustrated in Fig. 9-3 is 5.0 cm. For the detector and laser beam described above, determine the maximum angle θ that can be measured with this system in degrees and radians.

9-2. A Gaussian laser beam with a peak irradiance of I_0 and radius w is used to illuminate one quadrant of a quadrant detector as illustrated in Fig. 9-19.

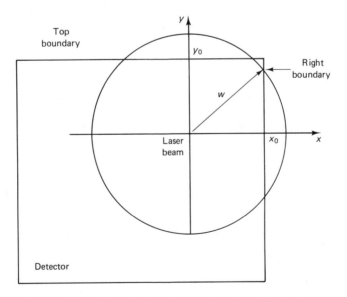

Figure 9-19 Cross section of Gaussian laser beam with radius w illuminating one quadrant of a gradrant detector.

(a) Determine the expression for the percentage of total beam power incident on a quadrant in terms of error functions and the position of quadrant boundaries x_0 and y_0. The total beam power in terms of peak irradiance is given by

$$P_{total} = \frac{1}{2}\pi w^2 I_0$$

and

$$\text{erf}(s) = \frac{1}{\sqrt{2\pi}}\int_{-\infty}^{s} e^{-v^2/2}\,dv$$

where $\text{erf}(s)$ is the error function.

(b) Assume that $x_0 = y_0$ and plot a graph of percentage of beam power incident on a quadrant for the interval $0 \le x_0 \le 1.4w$. Tables of error functions can be found in most mathematical handbooks.

9-3. Equation (9-1) can be derived using matrix methods for paraxial optics.

(a) Determine the transform matrices for both lens systems illustrated in Fig. 9-3.

(b) Determine a column vector that can be used to represent the center line of the laser beam incident on the alignment optics illustrated in Fig. 9-3.

(c) Use the results of parts (a) and (b) to derive Eq. (9-1) for both optical systems.

9-4. Using $y = F\theta$, $y = F\sin\theta$, and $y = 2F\tan(\theta/2)$ and assuming that $F = 20$ cm:

(a) Plot beam displacement versus angular displacement for $0 \le \theta \le \pi/6$ on a single graph.

(b) Find the beam displacement variation from $y = F\theta$ for $y = F\sin\theta$ and $y = 2F\tan(\theta/2)$ at $\theta = 3°$, $6°$, $12°$, $20°$, and $30°$.

9-5. Using $v = 2\omega F$, $v = 2\omega F\cos\theta$, and $v = 2\omega F\sec^2(\theta/2)$ and assuming that $F = 20$ cm and $\omega = 2000\pi$ rad/s:

(a) Plot beam velocity versus angular displacement for $0 \le \theta \le \pi/6$ on a single graph.

(b) Find the beam velocity variation from $v = 2\omega F$ for $v = 2\omega F\cos\theta$ and $v = 2\omega F\sec^2(\theta/2)$ at $\theta = 3°$, $6°$, $12°$, $20°$, and $30°$.

9-6. The focal length of the first converging lens in a scanner system like the one illustrated in Fig. 9-5 is 10 cm. A photodiode is used to trigger a timer circuit at $\theta_0 = \pi/6$ when time is considered to be zero, and the mirror rotates with an angular velocity of 120π rad/s.

(a) Use Eq. (9-4) to find the diameter of the object being measured if the laser beam scans past the top and bottom object edge at $t_1 = 0.33$ ms and $t_2 = 1.00$ ms.

(b) Use Eq. (9-7) to find the rod diameter.

(c) Which answer is more accurate? Why?

9-7. The HeNe laser beam used in the laser scanner system discussed in Problem 9-6 has a 1.0-mm radius waist that is located 20 cm in front of the first converging lens.

(a) Use the equations derived by Self to determine the location and radius of the waist on the output side of the first lens.

(b) Determine the time required for this waist to scan past an object edge.

(c) Repeat parts (a) and (b) for an input laser beam waist radius of 3.0 mm.

9-8. The equation of a parabola is

$$z = \frac{y^2}{4F}$$

where F is the focal length. Assume that a laser beam originates from the focal point of a parabolic mirror at an angle θ with respect to the optical axis. Show that the laser beam is reflected from the mirror so that its displacement y from the optical axis is

$$y = 2F \tan\left(\frac{\theta}{2}\right)$$

9-9. Using Eq. (9-13), derive

$$\Delta z' = m^2 \frac{\cos\theta}{\cos\phi} \Delta z = \frac{F^2 \cos\theta}{(o - F)^2 \cos\phi} \Delta z$$

9-10. Using Eq. (9-12) and (9-13), show that

$$\Delta z' = m^2 \Delta z$$

as θ approaches zero and that

$$\Delta z' = |m| \Delta z$$

for $\theta = \pi/2$.

9-11. Derive Eqs. (9-14) and (9-15) using the Scheimpflug optical design condition.

9-12. In many computer vision systems used in a line-of-light triangulation gage, the detector is fixed at an angle $\phi = \pi/2$.

(a) Under what condition will the line-of-light image be in best focus on the detector?

(b) Derive the equations for $\Delta z'$ and $\Delta x'$ for this condition.

9-13. For the line-of-light triangulation system illustrated in Fig. 9-13, assume that $\beta = 45°$, $\theta = 0°$, $F = 55$ mm, and the object distance for the lens is 20 cm. Assume the detector is a 512 by 512 photodiode array with photodiodes placed on 25-μm centers. If the vision system indicates that $\Delta z'$ and $\Delta x'$ are 250 and 100 detector spaces, respectively, find the:

(a) Image distance required for this application.

(b) Magnification.

(c) Angle ϕ required to maintain best image focus.

(d) Width and depth of the groove.

(e) Using the distance between two photodiodes as a least count, determine the minimum value for Δz and Δx that can be measured with this system.

9-14. An optical radar system is to be designed with a depth of field of 20 cm.
   (a) Determine the modulation frequency f_m required for this application.
   (b) Determine the range resolution of this system if the phase detector used is able to measure a phase difference of $\Delta\phi = 0.0032$ rad.

9-15. A diode-array camera gage is configured as shown in Fig. 9-16. A 1024 linear array with photodiodes placed at 25-μm centers is used as a detector. Using the Newtonian form of the lens equation, determine the following for $z = 62.5$ mm and $z' = 48.4$ mm.
   (a) Focal length of the lens.
   (b) Magnification.
   (c) Number of diodes that would not be illuminated if the gage is used to measure the diameter of a quarter-inch rod.
   (d) Theoretical resolution of the gage.

REFERENCES

BALLARD, D. H., AND C. M. BROWN, *Computer Vision.* Englewood Cliffs, NJ: Prentice Hall, 1982.

GOODMAN, J. W., *Introduction to Fourier Optics.* New York: McGraw-Hill Book Co., 1968.

HECHT, E., *Optics.* Reading, MA: Addison-Wesley Publishing Co., 1987.

JENKINS, F. A., AND H. E. WHITE, *Fundamentals of Optics.* New York: McGraw-Hill Book Co., 1976.

KINGSLAKE, R., *Optical System Design.* New York: Academic Press, Inc., 1983.

PEDROTTI, F. L., AND L. S. PEDROTTI, *Introduction to Optics.* Englewood Cliffs, NJ: Prentice Hall, 1987.

WILSON, J., AND J. F. B. HAWKES, *Optoelectronics, An Introduction.* Englewood Cliffs, NJ: Prentice Hall, 1989.

Chapter 10

Introduction
to Interferometry

The purpose of this chapter is to review the concept of interferometry and introduce the reader to several industrial applications that use this concept. This chapter will also provide background material for the holographic and speckle interferometry discussed in Chapter 11.

10-1 MATHEMATICAL DESCRIPTIONS OF LASER LIGHT

One solution to the wave equation, Eq. (1-1), is the equation of a plane wave propagating along the z axis:

$$\mathbf{E} = \hat{\mathbf{i}} \, E_0 \cos(kz - \omega t + \phi_0) \qquad (10\text{-}1)$$

where $\mathbf{E}$ is the electric field; E_0 is the electric field amplitude; $k = 2\pi/\lambda$, the propagation constant; $\omega = 2\pi f$, the angular frequency; and ϕ_0 is the original phase angle. The wavelength of light is λ and f is its frequency. A unit vector $\hat{\mathbf{i}}$ can be used in the equation to represent linearly polarized light. Irradiance I can be found by using another version of Eq. (1-5):

$$I = \frac{\langle \mathbf{E} \cdot \mathbf{E} \rangle}{Z} \qquad (10\text{-}2)$$

Recall that Z is the intrinsic impedance of the medium through which the light is propagating. The angular brackets $\langle \ \rangle$ denote the time average

$$\langle f(t) \rangle = \frac{1}{T} \int_0^T f(t) \, dt \tag{10-3}$$

where T is the response time of the equipment that measures the irradiance or the exposure time for the photosensitive emulsion that records an image.

The frequency of light in the visible regime is on the order of 10^{14} Hz. Light detectors available today cannot respond to this frequency and measure an average value. Thus the time-dependent portion of Eq. (10-1) is not required to describe coherent light. To facilitate separation of the time-dependent term, it is convenient to use complex notation. Euler's identities state that

$$e^{+j\beta} = \cos \beta + j \sin \beta$$

and

$$e^{-j\beta} = \cos \beta - j \sin \beta$$

where $j = \sqrt{-1}$. Using the first identity, Eq. (10-1) can be expressed as

$$\mathbf{E} = \{\text{Re}\} \, \hat{i} E_0 e^{+j(kz - \omega t + \phi_0)} \tag{10-4}$$

where (Re) indicates that only the cos or real part of the Euler identity is used. Note that the second identity could also be used. It is common practice, in the study of coherent optics, to drop the {Re} notation, and it is understood that only the real part of the Euler identity is used. Separation of the time-dependent term in Eq. (10-4) can be accomplished by rewriting the equation in the form

$$\mathbf{E} = \hat{i} E_0 e^{+j(kz + \phi_0)} \, e^{-j\omega t} \tag{10-5}$$

The time-independent term is called the *complex amplitude vector* and is given by

$$\mathbf{U} = \hat{i} E_0 e^{\pm j(kz + \phi_0)} \tag{10-6}$$

Since the time-dependent part of Eq. (10-5) is not required to describe coherent light, the complex amplitude vector given by Eq. (10-6) is sufficient to describe coherent light that acts as a plane wave. In Eq. (10-6) we have also added a $\pm$ sign in front of the argument $(kz + \phi_0)$. The positive sign can be used to represent a plane wave traveling in the $+z$ direction and the negative sign for a plane wave traveling in the opposite direction (see Problem 10-4). Other time-dependent mathematical descriptions, used in standard optics textbooks, can be expressed in terms of complex amplitude vectors and are summarized in Fig. 10-1. Recall that the propagation constant vector $\mathbf{k}$ has a magnitude of $2\pi/\lambda$ and is in the propagation direction. Light emanating from or converging to a point has spherical wavefronts, and the elec-

Plane polarized
plane wave traveling
in +z direction

$$\mathbf{U} = \hat{\mathbf{i}}\, E_0 e^{j(kz + \phi_0)}$$

Right-circularly
polarized traveling wave

$$\mathbf{U} = (\hat{\mathbf{i}} - j\hat{\mathbf{j}})E_0 e^{j(kz + \phi_0)}$$

Left-circularly
polarized traveling wave

$$\mathbf{U} = (\hat{\mathbf{i}} + j\hat{\mathbf{j}})E_0 e^{j(kz + \phi_0)}$$

Elliptically polarized
traveling wave

$$\mathbf{U} = \hat{\mathbf{i}}\, E_{0x} e^{jkz} + \hat{\mathbf{j}}\, E_{0y} e^{j(kz + \delta)}$$

Plane polarized
plane wave traveling
in arbitrary direction

$$\mathbf{U} = \hat{\mathbf{p}}\, E_0 e^{j(\mathbf{k}\cdot\mathbf{r} + \phi_0)}$$

Polarized coherent
light from point source
at origin

$$\mathbf{U} = \hat{\mathbf{e}}\, \frac{A}{r}\, e^{j(kr + \phi_0)}$$

Figure 10-1 Complex amplitude vectors.

tric field amplitude varies inversely proportional to r, and A is the amplitude a unit distance from the point.

Complex amplitude vectors also provide a more convenient way to determine the irradiance of coherent light. For all the mathematical descriptions,

$$\mathbf{E} = \frac{1}{2} [\mathbf{U}e^{-j\omega t} + \mathbf{U}^* e^{+j\omega t}]$$

where the symbol * indicates complex conjugation. The dot product

$$\mathbf{E} \cdot \mathbf{E} = \frac{1}{4} [2\mathbf{U} \cdot \mathbf{U}^* + \mathbf{U}^2 e^{-2j\omega t} + \mathbf{U}^{*2} e^{+2j\omega t}]$$

and

$$\langle \mathbf{E} \cdot \mathbf{E} \rangle = \frac{1}{2} \mathbf{U} \cdot \mathbf{U}^*$$

Time averages of the time-dependent terms are equal to zero due to the fact that the period T over which they are averaged is much greater than the electric field period (see Problem 10-5). Thus Eq. (10-2) becomes

$$I = \frac{\mathbf{U} \cdot \mathbf{U}^*}{2Z} \tag{10-7}$$

We will use this equation extensively to determine the irradiance due to interfering laser beams.

10-2 LASER DOPPLER VELOCIMETRY

Light reflected from a moving object is frequency shifted. If this light is mixed with light of a different frequency, the light will interfere and beat. This beat frequency can be used to determine the object's velocity. The concept of doppler frequency shifting will be reviewed and heterodyne detection discussed. This review and discussion are followed by examples of velocity measurement techniques that use these concepts.

Doppler frequency shift of reflected light.   Doppler shifting of reflected light from a pressure wave in an acoustic-optical modulator is discussed in Chapter 1. Equation (1-44) can be modified and rewritten as

$$\Delta f = 2 \frac{v_B}{\lambda} \cos(\psi/2) \tag{10-8}$$

where Δf is the doppler frequency shift of the reflected light and ψ is the angle between the light incident and reflected from the object. The velocity

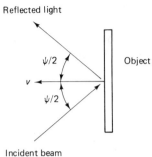

Reflected light

$\psi/2$

v

$\psi/2$

Incident beam

Object

Figure 10-2 System geometry for laser light reflected from moving object, where velocity component v is parallel to bisector of angle ψ between incident and reflected light.

component v_B is parallel to the bisector of the angle ψ. This geometry is illustrated in Fig. 10-2.

Heterodyne detection.   As illustrated in Fig. 10-3, assume that two laser beams with slightly different frequencies are superimposed and incident on a photodetector. Assume the electric field intensity vectors are given by

$$\mathbf{E}_1 = \hat{\mathbf{i}} E_{01} \cos(k_1 z - \omega_1 t + \phi_{01}) \qquad (10\text{-}9)$$

and

$$\mathbf{E}_2 = \hat{\mathbf{i}} E_{02} \cos(k_2 z - \omega_2 t + \phi_{02}) \qquad (10\text{-}10)$$

Assuming that the beams are superimposed at $z = 0$ and that the original phase angles are equal to zero. The resultant electric field vector is

$$\mathbf{E} = \hat{\mathbf{i}}[E_{01} \cos(\omega t) + E_{02} \cos(\omega + \Delta\omega)t] \qquad (10\text{-}11)$$

where $\omega_1 = \omega$ and $\omega_2 = \omega + \Delta\omega$.

$$\langle \mathbf{E} \cdot \mathbf{E} \rangle = E_{01}^2 \langle \cos^2 \omega t \rangle + E_{02}^2 \langle \cos^2(\omega + \Delta\omega)t \rangle$$
$$+ 2E_{01}E_{02} \langle \cos(\omega t) \cos(\omega + \Delta\omega)t \rangle \qquad (10\text{-}12)$$

The average value of cosine squared terms is one-half, and

$$\langle \cos(\omega t) \cos(\omega + \Delta\omega)t \rangle = \frac{1}{2} \langle \cos(\Delta\omega\, t) + \cos(2\omega + \Delta\omega)t \rangle$$

The resultant irradiance is

$$I = I_1 + I_2 + \sqrt{I_1 I_2} \langle \cos(\Delta\omega\, t) + \cos(2\omega + \Delta\omega)t \rangle \qquad (10\text{-}13)$$

where $I_1 = E_{01}^2/2Z$ and $I_2 = E_{02}^2/2Z$. The $\cos(\Delta\omega\, t)$ term oscillates much more slowly than the other cosine term. By using a bandpass filter that only passes information about the $\cos(\Delta\omega\, t$ term, the output of the filter can be expressed as

$$O = R\,[\sqrt{I_1 I_2}\,\cos(\Delta\omega\, t)] \qquad (10\text{-}14)$$

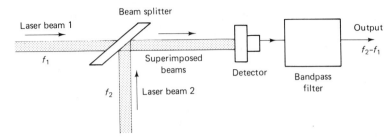

Figure 10-3 Schematic of heterodyne detector measuring beat frequency of superimposed laser beams of two different frequencies.

where R depends on the bandpass filter characteristics. Thus the output from the filter is sinusoidal and has angular frequency $\Delta\omega = \omega_2 - \omega_1 = 2\pi(f_2 - f_1) = 2\pi f_B$. The difference in frequencies f_B is the *beat frequency*.

Reference beam technique.   Figure 10-4 is a simplified diagram of a laser doppler vibrometer designed to measure vibrating surface velocity. Laser light is focused onto a surface using a converging lens. Light reflected from the surface passes back through the converging lens. Light reflected at an angle ψ relative to the incident light is reflected by mirror 2, passes through beam splitter 2, and is incident on a photodetector. We will refer to this light as the *object beam*. A portion of the light from the laser is reflected by beam splitter 1, reflected by mirror 1, and reflected by beam splitter 2 to the detector. This beam is referred to as the *reference beam*. At the detector the reference beam is mixed with the object beam. These two beams beat, and the beat frequency is detected by a heterodyne detector.

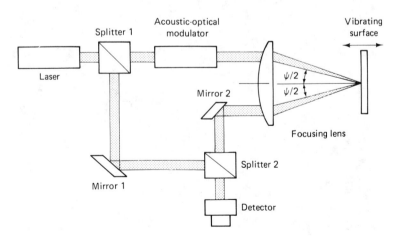

Figure 10-4 Simplified diagram of laser doppler system that uses the reference-beam technique to measure the velocity of a vibrating object.

The magnitude of the velocity component v parallel to the bisector of angle ψ can be determined, but not its direction. Direction can be determined by frequency shifting the light incident on the object surface. In our example we are using an acoustic-optical modulator. Using this device, the beat frequency f_B is

$$f_B = 2 \frac{v}{\lambda} \cos\left(\frac{\psi}{2}\right) + \Delta f_a \qquad (10\text{-}15)$$

where Δf_a is the frequency shift due to the acoustic-optical modulator. Thus the velocity component v parallel to the bisector of the angle ψ is

$$v_B = \frac{\lambda(f_B - \Delta f_a)}{2 \cos(\psi/2)} \qquad (10\text{-}16)$$

If velocity component v_B is negative, the object is moving away from the laser doppler vibrometer, and if positive, it is moving toward the vibrometer. By integrating and differentiating the velocity component, displacement and acceleration can be determined. A photograph of a system available from a vendor is shown in Fig. 10-5.

The configuration discussed above is commonly referred to as using the *reference-beam technique*. An alternative is to use two coherent object beams incident onto a surface from two different angles. We will refer to this technique as the *dual-object-beam technique*. In this technique the two object beams reflected by a moving object are doppler shifted by two dif-

Figure 10-5 Laser doppler vibrometer being used to analyze hard disk drive. (Courtesy of DANTEC.)

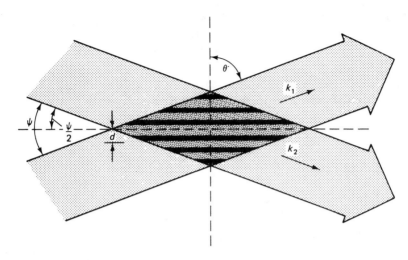

Figure 10-6 Intersecting coherent laser beams.

ferent amounts. Thus, when superimposed, the reflected light from a moving object beats and can be used to determine velocity. Equations for the dual-object-beam technique can be derived using the doppler effect or interference. We have chosen to use interference to derive these equations. Students will use the doppler effect to derive these equations in Problem 10-9.

Dual-object-beam technique.   Consider two coherent plane waves with the same amplitude and linear polarization that intersect at an angle ψ, as shown in Fig. 10-6. In the region of intersection, bright and dark interference planes are produced that are parallel to the plane bisecting the angle between the two beams. To derive the expression for the irradiance along the x axis, we will use the equation for a polarized plane wave propagating in the **k** direction from Fig. 10-1. By assuming that $r = x$ and $\phi_0 = 0$, this equation can be expressed as

$$\mathbf{U} = \hat{\mathbf{j}}E_0 e^{jkx \cos \theta}$$

Thus the complex amplitude vectors of the two waves can be described by

$$\mathbf{U}_1 = \hat{\mathbf{j}}E_0 e^{jkx \sin(\psi/2)} \tag{10-17}$$

and

$$\mathbf{U}_2 = \hat{\mathbf{j}}E_0 e^{-jkx \sin(\psi/2)} \tag{10-18}$$

The resultant complex amplitude due to the interference of the two plane waves is

$$\mathbf{U} = \hat{\mathbf{j}}E_0 [e^{jkx \sin(\psi/2)} + e^{-jkx \sin(\psi/2)}]$$

Using Eq. (10-7), the irradiance I is given by

$$I = 4I_0 \cos^2 \left[kx \sin \left(\frac{\psi}{2} \right) \right] \tag{10-19}$$

where $I_0 = E_0^2/2Z$ and is the irradiance of either of the intersecting beams. The distance d between two adjacent bright fringes can be found by letting

$$kd \sin \left(\frac{\psi}{2} \right) = \pi$$

Thus

$$d = \frac{\lambda}{2 \sin(\psi/2)} \tag{10-20}$$

Not only is this equation useful to describe laser doppler dual-beam velocimetry, but it can be used to describe the spacing of fringes recorded in holography. If ψ is small, interference fringes can be recorded on low-resolution photographic emulsion, and if ψ is large (at or near 180°), an extremely high resolution emulsion is required.

Figure 10-7 is a simplified drawing of a system designed to determine the velocity of particles moving in a transparent gas or liquid. In this system a laser beam is first split into two beams. These beams pass through two holes in a mirror. Then they pass through a converging lens that focuses the beams to its focal plane, where they intersect, producing interference planes. If the objective is to determine the flow velocity of a gas or liquid, particles with the same mass density as the medium are frequently used to seed the medium. As a particle passes through these planes, it scatters light. The light

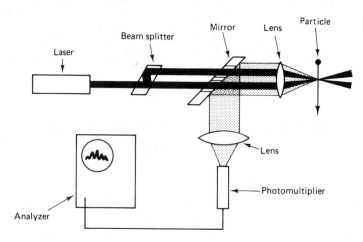

Figure 10-7   Schematic of laser velocimetry system that uses the dual-object technique to measure particle velocity.

scattered from the particle blinks as it passes through the bright and dark planes. For the system illustrated, a portion of the scattered light passes back through the converging lens. A portion of this light is reflected to a second converging lens that focuses this light to a detector. The detector used could be a photomultiplier or an avalanche photodiode. The scattered light and thus the signal from the detector are periodic and have a frequency of

$$f = \frac{v_x}{d} = \frac{2v_x \, \sin(\psi/2)}{\lambda} \tag{10-21}$$

where v_x is the x component of the particle velocity that is perpendicular to the interference planes. Other components of the velocity can be found by forming fringes that are orthogonal to the first set. Two orthogonal fringe sets can be formed by passing four beams through a single converging lens. To produce three orthogonal fringe sets would require additional optics and another pair of laser beams incident from a direction perpendicular to the other two pairs. Signal separation for each fringe set can be accomplished by using laser light of different colors.

Due to its noncontact nature, potential applications of laser doppler velocimetry to measure particle velocity and analyze vibration in industry seem vast. Using fiber optics to deliver and collect light should allow this technique to be used in confined spaces. We will also see increased use of laser doppler velocimeters in other applications, such as measuring the speed of cars on the highway.

10-3 MICHELSON INTERFEROMETER

The Michelson interferometer illustrated in Fig. 1-24 can be used to make precise linear displacement measurements of one mirror relative to another or to detect and/or measure surface variations between two mirror surfaces. We will first consider a Michelson interferometer system, as shown in Fig. 10-8, that uses a diverging lens to produce a diverging spherical wave. The beam splitter splits the diverging beam into two beams of equal irradiance. The two beams are reflected by corresponding mirrors, and equal portions are recombined by the beam splitter. These recombined waves interfere, producing fringes on a frosted glass screen that can be observed with the human eye.

The diverging wave can be described by the complex amplitude equation, from Fig. 10-1, for a spherical wave emanating from a point source:

$$U = \frac{A}{r} \, e^{j(kr + \phi_0)} \tag{10-22}$$

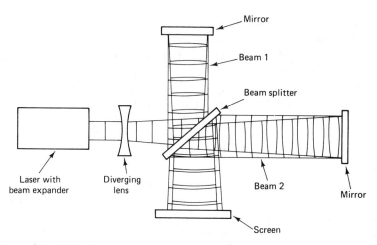

Figure 10-8   Michelson interferometer using diverging laser beam as light source.

Wavefront radius r can be expressed in terms of coordinates z and ρ. As shown in Fig. 10-9, z is the distance from the spherical wave point source to the frosted glass screen, and ρ is the radial distance from the z axis to an observation point on the screen. Wavefront radius at the screen is

$$r = \sqrt{z^2 + \rho^2} = z \sqrt{1 + \frac{\rho^2}{z^2}}$$

By considering only paraxial rays, we can consider ρ to be sufficiently small so that

$$z \sqrt{1 + \frac{\rho^2}{z^2}} \simeq z\left(1 + \frac{\rho^2}{2z^2}\right) = z + \frac{\rho^2}{2z}$$

Equation (10-22) can be rewritten as

$$U \simeq \frac{A}{r} \, e^{j(kz + k\rho^2/2z + \phi_0)} \tag{10-23}$$

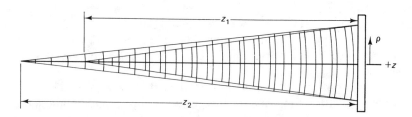

Figure 10-9   Superposition of two spherical waves with different radii propagating in the same direction.

If we assume that amplitudes of two beams at the screen are equal and the original phase angles are zero, beam 1 can be described by

$$U_1 = E_0 e^{jk(z_1 + \rho^2/2z_1)}$$

and beam 2 by

$$U_2 = E_0 e^{jk(z_2 + \rho^2/2z_2)}$$

The resultant irradiance of the two interfering beams is

$$I = 4I_0 \cos^2\left(\frac{k}{2}\right)\left[\Delta z + \frac{\rho^2}{2}\left(\frac{1}{z_1} - \frac{1}{z_2}\right)\right] \qquad (10\text{-}24)$$

where $\Delta z = z_2 - z_1$ and $I_0 = E_0^2/2Z$. A photograph of the interference pattern produced is shown in Fig. 10-10. By observing the center of the pattern at $\rho = 0$, we can reduce Eq. (10-24) to

$$I = 4I_0 \cos^2\left(\frac{k\,\Delta z}{2}\right) \qquad (10\text{-}25)$$

Bright fringes are observed at $\rho = 0$ if

$$\frac{k\,\Delta z}{2} = m_B \pi$$

or

$$\Delta z = m_B \lambda \qquad (10\text{-}26)$$

where $m_B = 0, 1, 2, 3, \ldots$. Displacing one mirror of the Michelson interferometer a half-wavelength in the propagation direction of the light changes

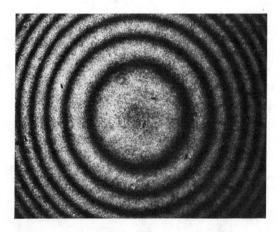

Figure 10-10 Photograph of the interference pattern due to two spherical waves with different radii propagating in the same direction.

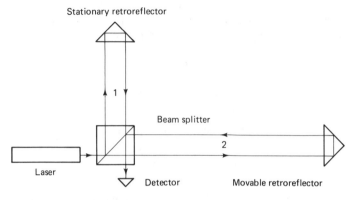

Figure 10-11 Simplified diagram of Michelson interferometer designed to measure linear displacement.

the optical path length of that light by one wavelength. Thus, at $\rho = 0$, if the interference pattern is bright, it will change to dark and then back to bright as the mirror is moved one half-wavelength. Dark fringes can be represented by half-integers $m_D = \frac{1}{2}, \frac{3}{2}, \frac{5}{2}, \ldots$. Thus, at $\rho = 0$, if the pattern is bright, it will become dark when one mirror is moved one quarter-wave-

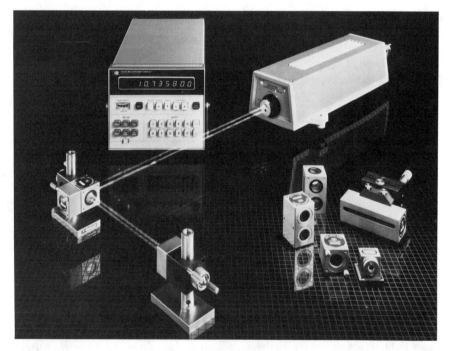

Figure 10-12 Two-frequency laser measurement system. (Courtesy of Hewlett-Packard.)

Figure 10-13   Interference fringes due to the superposition of two collimated coherent laser beams intersecting at a small angle.

length. Thus the Michelson interferometer can be used as a displacement measurement device that has a least count of one quarter-wavelength.

Commercial interferometers based on the principles of the Michelson interferometer are available. A simplified schematic for systems of this type is shown in Fig. 10-11. The flat mirrors are replaced by retroreflectors. Retroreflectors do not require the precise alignment required for flat mirrors. The system illustrated by Fig. 10-11 can determine displacement magnitude but not direction.

The commercial laser measurement system shown in Fig. 10-12 uses a two-frequency laser to determine displacement direction. Interferometers

Figure 10-14   Interference fringes due to a small bump on one mirror of a Michelson interferometer.

of this type can be used to measure linear displacement, velocity, flatness, pitch, yaw, straightness, squareness, parallelism, and perpendicularity. They also can make measurements with an accuracy of less than a quarter-wavelength. These systems are usually interfaced with a computer to compensate for changes of optical path length of the light through air due to temperature, humidity, and pressure.

To check surface variation between two mirrors, we consider using

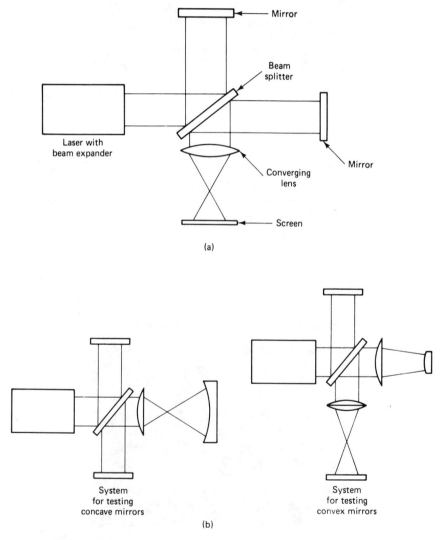

(a)

(b)

Figure 10-15 Schematics of (a) Twyman–Green interferometer and (b) laser interferometers designed to test concave and convex mirrors.

the Michelson interferometer illustrated in Fig. 10-8 with the diverging lens removed. If both mirrors are optically flat and one mirror is rotated through a small angle, straight, equally spaced fringes result, as shown in Fig. 10-13. Surface variations of one mirror are indicated by variation in the straightness and spacing of these lines. Figure 10-14 is a photograph showing the variation of these fringes due to a small bump on one mirror of a Michelson interferometer. Variation of these fringes can also be due to variation in the optical path length of the light between the beam splitter and one of the mirrors. Change in optical path length can be due to introducing a different medium or changing the refractive index of the medium through which the light travels. Refractive index change can be due to change in temperature and pressure.

So that an observer can see the surface being tested as well as the interference fringes, an imaging lens is often added to a Michelson interferometer, as illustrated in Fig. 10-15(a). Using this technique, the image of the surface tested and the interference fringes are superimposed. This image can be projected on a screen, photographic film, or the detector of a TV camera. Interferometers of this type are called Twyman–Green laser interferometers and are often designed to test curved surfaces, as well as flat surfaces as illustrated in Fig. 10-15(b).

10-4 FIZEAU INTERFEROMETER

A Fizeau interferometer is illustrated in Fig. 10-16. We will consider when an optical flat is used as a reference surface and the surface to be tested is a convex mirror. The reflection from the reference surface is a plane wave, and the reflection from the test surface is a diverging spherical wave. The plane wave can be described by Eq. (10-6) and the diverging spherical wave by Eq. (10-23). By using an imaging lens to observe the convex mirror surface as well as the interference pattern produced, we are in effect observing the apparent interference of the plane wave and spherical wave at convex mirror surface. By setting the amplitudes of the two waves equal at this location and assuming the initial phase angle of the spherical to be zero, the plane wave can be described by

$$U_1 = E_0 e^{j(kz_1 + \phi_0)}$$

and the spherical wave by

$$U_2 = E_0 e^{jk(z_2 + \rho^2/2z_2)}$$

Distance z_2 is the wavefront radius of the spherical wave at the vertex of the convex mirror, or the mirror focal length. Radius ρ is the radial distance

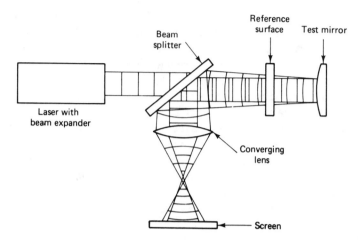

Figure 10-16 Schematic of a Fizeau laser interferometer.

from the optical axis to an observation point on the convex mirror surface. Irradiance of the two superimposed waves is

$$I = 4I_0 \cos^2 \left[\frac{1}{2} \left(\frac{k\rho^2}{2z_2^2} + k \, \Delta z - \phi_0 \right) \right] \tag{10-27}$$

If the center of the observed interference is bright, then $k \, \Delta z - \phi_0 = 2m_0\pi$, where m_0 is the order of the central bright fringe. Other bright fringes accrue when

$$\frac{k\rho^2}{2z_2^2} = 2m_B \pi$$

where $m_B = 0, 1, 2, 3, \ldots$. Since z_2 is the focal length of the convex mirror, the radius of curvature of this mirror $R = 2z_2$. Therefore,

$$R = \frac{\rho^2}{m_B \lambda} \tag{10-28}$$

This equation is very similar to the equation that is used to find the radius of Newton rings, which is to be expected, because the only difference between the Fizeau interferometer and a Newton ring apparatus is that the reference flat and surface to be tested are separated by some distance, rather than being in intimate contact. Also, it should be realized that Eq. (10-28) is an approximation assuming that $R \gg \rho$.

As in the case of the Twyman–Green interferometer, the Fizeau interferometer can be configured to evaluate flat as well as curved surfaces. Two-dimensional diode arrays interfaced with a computer can interpret the interference fringes produced and quantify the results. A photograph of a commercially available Fizeau interferometer is shown in Fig. 10-17.

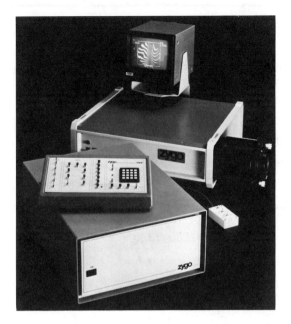

Figure 10-17   Fizeau laser interferometer system. (Courtesy of Zygo.)

PROBLEMS

10-1. Given the expression for the electric field

$$\mathbf{E} = \hat{\mathbf{i}}\,\frac{10\ \mu V}{m}\,\cos\left(\frac{10^{7}}{m}\,z + \frac{3\times10^{15}}{s}\,t + 0.75\right)$$

find the:
(a) Amplitude.
(b) Propagation constant.
(c) Wavelength.
(d) Angular frequency.
(e) Frequency.
(f) Original phase angle.
(g) Irradiance in free space.

10-2. Given the expression for the complex amplitude of the electric field due to a point source

$$U = \frac{100\ \mu V}{r}\,\cos(\psi)\ e^{j(1.22\times10^{7}/m)r}$$

find the:
(a) Amplitude.
(b) Propagation constant.
(c) Wavelength.

(d) Frequency.

(e) Original phase angle.

(f) Irradiance at $\psi = 15°$ and $r = 1.0$ m.

10-3. By taking the first differential of the cosine function argument in Eq. (10-1), show that the traveling plane wave is propagating in the $+z$ direction.

10-4. Given the expression for the complex amplitude

$$U = E_0 e^{-j(kz + \phi_0)}$$

find the direction of propagation of $E_1 = U e^{-j\omega t}$ and $E_2 = U^* e^{-j\omega t}$.

10-5. Show that the time average of

$$U^2 e^{-j2\omega t}$$

approaches zero for exposure time T much greater than the period of electric field oscillation due to coherent light.

10-6. A sinusoidal plane wave is described by

$$U = E_0 e^{j(\mathbf{k} \cdot \mathbf{r} + \phi_0)}$$

By finding the appropriate expressions of $\mathbf{k} \cdot \mathbf{r}$, write equations that describe a wave in terms of wavelength and the rectangular coordinates x, y, and z, if the wave is propagating:

(a) Along the $+z$ axis.

(b) Along the line $x = z$, $y = 0$.

(c) Perpendicular to the planes $x + y + z = $ constant.

10-7. At time zero the equation for a traveling plane wave is

$$E = E_0 e^{j(\mathbf{k} \cdot \mathbf{r})}$$

If the wave is traveling at 45° to the z axis, show that the position along the z axis where $E = E_0$ is

$$z = m\lambda\sqrt{2}, \quad \text{where } m = 0, \pm1, \pm2, \pm3 \ldots$$

10-8. Signal output from a laser doppler vibrometer with the system geometry shown in Fig. 10-4 is illustrated in Fig. 10-18. If the system uses a HeNe laser and $\psi = 10°$, determine the following:

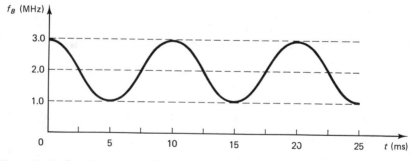

Figure 10-18   Signal output, beat frequency versus time, from a laser doppler vibrometer.

(a) Vibration frequency.

(b) Peak and rms velocity.

(c) Peak and rms acceleration.

(d) Vibration amplitude.

10-9. Derive Eq. (10-21) using the doppler effect.

10-10. Determine the maximum and minimum distance between adjacent fringes due to two interfering laser beams that can be described as coherent plane waves.

10-11. Derive the expression for the irradiance I as a function z due to two interfering plane waves

$$\mathbf{U}_1 = \hat{\mathbf{i}}E_0 e^{jkz}$$

and

$$\mathbf{U}_2 = \hat{\mathbf{i}}E_0 e^{-jkz}$$

Plot irradiance I as a function of z and determine the distance between adjacent bright fringes. Repeat the exercise for the two plane waves

$$\mathbf{U}_1 = \hat{\mathbf{i}}E_0 e^{jkz}$$

and

$$\mathbf{U}_2 = \hat{\mathbf{j}}E_0 e^{-jkz}$$

10-12. Derive the expression for the irradiance I as a function of x and y due to two interfering plane waves

$$\mathbf{U}_1 = \hat{\mathbf{i}}E_0 e^{jkz}$$

and

$$\mathbf{U}_2 = \hat{\mathbf{i}}E_0 e^{jky}$$

If the plane waves represent light from a HeNe laser, determine the distance between bright fringes and their spatial frequency (bright fringes per unit distance).

10-13. Derive the expression for the irradiance I as a function of ϕ_0 for an interfering linearly polarized beam and circularly polarized beam

$$\mathbf{U}_1 = \hat{\mathbf{i}}E_0 e^{j(kz + \phi_0)}$$

and

$$\mathbf{U}_2 = E_0(\hat{\mathbf{i}} + j\hat{\mathbf{j}})e^{jkz}$$

Plot irradiance I as a function of ϕ_0 and label its minimum, average, and maximum values.

10-14. Show that the expression for the radius of the dark fringes shown in Fig. 10-10 is

$$r = \sqrt{\frac{2z_1 z_2}{m_{0D} + \dfrac{1}{2}}} \sqrt{m_D}$$

where m_D and m_{0D} are the orders of the dark fringes and central dark fringe, respectively. Measure the radii of the dark fringes in Fig. 10-10. Use log–log paper or a computer curve-fitting program to show that the equation for the dark fringe radii have the form

$$r = am_D^b$$

Determine a and b.

REFERENCES

BALLARD, D. H., AND C. M. BROWN, *Computer Vision*. Englewood Cliffs, NJ: Prentice Hall, 1982.

HECHT, E., *Optics*. Reading, MA: Addison-Wesley Publishing Co., 1987.

JENKINS, F. A., AND H. E. WHITE, *Fundamentals of Optics*. New York: McGraw-Hill Book Co., 1976.

PEDROTTI, F. O., AND L. S. PEDROTTI, *Introduction to Optics*. Englewood Cliffs, NJ: Prentice Hall, 1987.

WILSON, J., AND J. F. B. HAWKES, *Optoelectronics, An Introduction*. Englewood Cliffs, NJ: Prentice Hall, 1989.

Chapter 11

Holographic and Speckle Interferometry

This chapter first provides a simplified description of holography. This is followed by mathematical descriptions of holography and holographic interferometry. Electronic speckle pattern interferometry is discussed as an alternative and extension of holographic interferometry. The resulting equations are then used to discuss engineering applications. The treatment of holographic and speckle interferometry in this text will be restricted to techniques we have found useful for solving engineering problems.

11-1 HOLOGRAPHY

Dennis Gabor published the first papers on holography in 1948. From 1948 through 1960, progress in the use and development of holography was slow, primarily because researchers did not have access to a light source that could produce intense light with a long coherence length. Emmett Leith and Juris Upatnieks, working at the University of Michigan Institute of Science and Technology in the early 1960s, produced the first holograms using a laser. In doing so, Leith and Upatnieks transformed holography from an obscure concept to a viable engineering tool. In 1965, Robert Powell and Karl Stetson published the first paper in holographic interferometry. Dennis Gabor received the Nobel prize in physics for his work in 1972.

For our purposes, a hologram can be thought of as an optical device, produced by using photographic techniques and laser light, that is capable of creating three-dimensional images. The word *hologram* stems from the Greek word *holos*, which means whole, complete, or entire, and *gram*, which means message. Thus a hologram is a complete record of an optical scene. In conventional photography the light reflected from a scene is focused, using a converging lens system, onto a photographic emulsion. Variation of the irradiance due to the image being focused on the emulsion is related only to the electric field amplitude of the light. In holography, a photosensitive medium is exposed to an interference pattern produced by two coherent laser beams. One beam, called the *object beam*, is reflected from a scene to the photosensitive medium. The other beam, called the *reference beam*, is reflected directly from the laser to the photosensitive medium using mirrors. Using this technique, both amplitude and phase information about the electric field due to the light reflected from the scene can be recorded. The photosensitive medium is exposed to both beams, and after development the resulting interference pattern is stored in the photosensitive material as a variation of transmittance, thickness, or refractive index. Using this technique, both phase and amplitude information due to light reflected from the object is captured and stored. The developed photosensitive medium with its recorded interference pattern is called a hologram. Depending on the type of hologram, the hologram can be played back using laser or white light to produce three-dimensional images.

Holography system.   Figure 11-1 is a sketch of an optical system that can be used to produce holograms. All the optical elements used in the production of optical holograms are mounted on an essentially vibration free surface. Usually, a heavy table with a steel or granite top, isolated from floor vibration by air tubes, provides this surface. A photograph of a holography system is shown in Fig. 11-2. A continuous or pulsed laser operating in the TEM_{00} mode at a wavelength within the visible portion of the electromagnetic spectrum is used for most engineering applications. Helium–neon, argon, and krypton are the continuous-wave lasers commonly used. Pulsed ruby lasers are used to produce holograms of transient phenomena or when vibration is a problem.

To produce a hologram, a laser beam is split into two beams by a beam splitter, as illustrated in Fig. 11-1. The reference beam is reflected by a mirror and then spread by a converging lens to uniformly illuminate a photosensitive medium. Photosensitive media frequently used in engineering applications of holography include photographic emulsions, photoresist, dicromated gelatin, and thermoplastic. Usually, the recording medium is laminated to a glass plate to eliminate vibration of the medium. The object beam is spread by a converging lens and then reflected to the object by a mirror so that the object is uniformly illuminated as observed from the position of the pho-

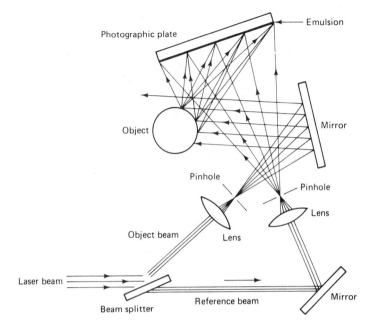

Figure 11-1   Basic optical system for producing holograms.

tosensitive medium. Pinholes used in the focal planes of the converging lenses act as spatial filters to remove optical noise from the laser light. The optical noise is primarily due to light scattered by dust on the lens surfaces. Maximum coherence of the object and reference beams at the photosensitive medium is obtained by making the object and reference beams equal in length.

It is desirable that the objects used in holography have diffusive surfaces. Flat white paint is often used on industrial parts to accomplish this.

Figure 11-2   Typical holography system.

When illuminated, each point on these surfaces will act as a point source, scattering light in all possible directions. Thus optical information about every illuminated point on the object can be recorded on an infinite number of points on the photosensitive medium.

Consider the complex amplitude of the laser light scattered from a point on the object to a point on the photosensitive medium to be U_O. The complex amplitude of the reference beam at the same point on the photosensitive medium is assumed to be U_R, as illustrated in Fig. 11-3. The resulting complex amplitude of the light at the point on the photosensitive medium is the vector sum

$$U = U_R + U_O \tag{11-1}$$

Using Eq. (10-7), the irradiance is found to be

$$I = \frac{1}{2Z} (U_R \cdot U_R^* + U_O \cdot U_O^* + U_R^* \cdot U_O + U_R \cdot U_O^*) \tag{11-2}$$

This result can be rewritten as

$$I = I_R + I_O + \frac{1}{2Z} (U_R^* \cdot U_O + U_R \cdot U_O^*) \tag{11-3}$$

where the first two terms are the irradiances of the reference and object beams, respectively. The third term depends on the relative phase and polarization of the interfering beams. If the object and reference beams are cross-polarized, $I = I_R + I_O$, all phase information is lost and a hologram cannot be produced. Ideally, then, both beams should be linearly polarized in the same direction. We will assume that this is the case in the remaining discussion of holography. Exposure $\mathscr{E}$ of the photosensitive medium is the product of the I and the exposure time T, or

$$\mathscr{E} = IT \tag{11-4}$$

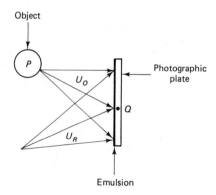

Figure 11-3   Recording geometry.

An absorption or phase hologram can be produced depending on the type of photosensitive medium used.

Absorption holograms.   An *absorption hologram* records the interference pattern due to the superposition of the reference and object beams as a transmittance variation of a photosensitive medium. Such is the case when a photographic emulsion is used. After a photographic emulsion is exposed to laser light in a holography system and then developed, it has a light amplitude transmittance versus exposure as illustrated in Fig. 11-4. To obtain a linear recording, the emulsion is biased by using the exposure due to the reference beam to establish a quiescent point in the linear portion of the transmittance versus exposure curve. At this quiescent point the transmittance is t_0 and the exposure is $I_R T$. Note that to accomplish this the irradiance of the reference beam must be greater than the irradiance due to the object beam. By making a linear recording, it can be assumed the amplitude transmittance of the photographic emulsion after development is a linear function of the irradiance incident on the emulsion during the exposure. The transmittance t can be expressed as

$$t = t_0 + \beta \left(I_O + \frac{U_R^* U_O}{2Z} + \frac{U_R U_O^*}{2Z} \right) \tag{11-5}$$

This is the equation of a straight line with t_0 being the vertical axis intercept and β a slope equal to the slope of the transmittance versus exposure curve at the quiescent point times the exposure time. Note that, if a photographic emulsion has a transmittance versus curve like the one illustrated in Fig. 11-4, it is a negative emulsion and β is negative.

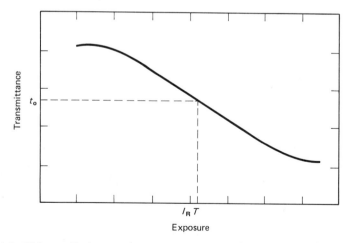

Figure 11-4 Light-amplitude transmittance versus exposure for typical photographic plate used for holography.

To reconstruct images due to a hologram after development, it is illuminated by a playback beam U_P. In most industrial applications a laser is used for reconstruction. The light transmitted through the hologram is

$$U_P t = U_P t_0 + \beta\left(U_P I_O + \frac{U_P U_R^* U_O}{2Z} + \frac{U_P U_R U_O^*}{2Z}\right) \qquad (11\text{-}6)$$

It is traditional to write Eq. (11-6) as

$$U_P t = U_1 + U_2 + U_3 + U_4 \qquad (11\text{-}7)$$

where

$$U_1 = U_P t_0$$

$$U_2 = \beta U_P I_O$$

$$U_3 = \frac{\beta U_P U_R^* U_O}{2Z}$$

$$U_4 = \frac{\beta U_P U_R U_O^*}{2Z}$$

It can be shown that the light is transmitted and diffracted by the hologram and is distributed as illustrated in Fig. 11-5. Complex amplitude U_1 represents light that is transmitted straight through the hologram. Light described by U_2 is diffracted by a small angle relative to U_1. This light is diffracted by low-frequency fringes, recorded on the hologram, that are due to interference of light scattered from different points on the object. Diffracted light described by U_3 and U_4 is capable of forming images.

For the system geometry discussed, a virtual image will be produced, as illustrated in Fig. 11-6, if

$$U_P = b U_R \qquad (11\text{-}8)$$

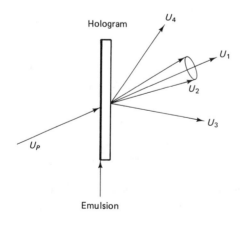

Figure 11-5 Light transmitted and diffracted by a hologram when illuminated with a playback beam.

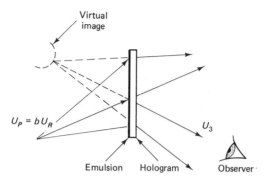

Figure 11-6 Virtual image reconstruction.

that is, if the playback beam is identical to the reference beam except for a multiplicative constant b. The diffracted light described by complex amplitude U_3 is

$$U_3 = (b\beta I_R)U_O \tag{11-9}$$

where $I_R = U_R U_R^*/2Z$ is the irradiance of the original reference beam. The light represented by U_3 is an exact duplicate of the original object beam except for the multiplicative constant $b\beta I_R$. Thus the light represented by this term is identical to the light reflected from the object to the photographic emulsion when it was exposed. An observer viewing the light described by U_3 will see a three-dimensional virtual image of the original object!

To reconstruct a real image,

$$U_P = bU_R^* \tag{11-10}$$

The reconstruction beam is the complex conjugate of the reference beam multiplied by a constant b. The complex conjugate of U_R represents a beam traveling in the opposite direction of the original reference beam, as illustrated in Fig. 11-7. Using this playback scheme,

$$U_4 = (b\beta I_R)U_O^* \tag{11-11}$$

Light represented by this term is an exact duplicate of the original object beam except for the multiplicative constant $b\beta I_R$ and the fact that the light is conjugated. This conjugated light travels in the opposite direction of the original object beam and produces a real image of the object.

Phase holograms.   A *phase hologram* records the interference pattern due to the superposition of the reference and object beam in a photosensitive medium as a variation in thickness or refractive index. One technique to produce a phase hologram is to produce an absorption hologram and then bleach it, removing the absorptive material from the hologram. Materials that can be used to produce a phase hologram directly include dichromated

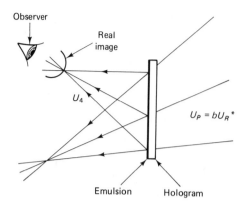

Figure 11-7   Real image reconstruction.

gelatin and thermoplastic. A phase hologram can diffract light more efficiently but is usually noisier than an absorption hologram.

To facilitate our discussion, we will assume that the holography system geometry is as illustrated in Fig. 11-3, and $U_R = A$ and $U_O = ae^{j\phi}$. Using Eq. (11-3), we find that the irradiance at a point on the photosensitive medium is

$$I = I_R + I_O + \frac{1}{2Z}(Aae^{j\phi} + Aae^{-j\phi})$$

or

$$I = I_R + I_O + \frac{Aa}{Z}\cos\phi \qquad (11\text{-}12)$$

After proper exposure to the interfering object and reference beams, the photosensitive medium is processed. We will assume that the interference pattern is linearly recorded on a thermoplastic plate as a variation in thickness, as illustrated in Fig. 11-8. After processing, the transmittance t of the phase hologram is

$$t = e^{j\Delta\psi} \qquad (11\text{-}13)$$

where $\Delta\psi$ is the phase shift due to a variation from the normal thickness of the thermoplastic plate ΔT. Note that Eq. (11-13) assumes that the light amplitude is not affected as it is transmitted through the phase hologram. The phase change $\Delta\psi$ is related to variation of thermoplastic thickness ΔT by

$$\Delta\psi = k(n - 1)\Delta T \qquad (11\text{-}14)$$

where n is the thermoplastic refractive index and k is the propagation constant. The variation of thermoplastic thickness ΔT is assumed to be a linear function of irradiance variation $\Delta T = -\gamma\Delta I$, where γ is a proportionality

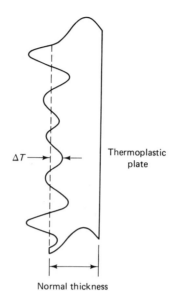

Figure 11-8 Variation of thermoplastic plate thickness due to holographic recording.

constant. Thermoplastic plates used for holography can only produce variation in thickness over a limited spatial frequency range, as shown in Fig. 11-9. Variations in thickness at low spatial frequencies due to I_R and I_O are not recorded. Thus the only term in Eq. (11-12) that represents irradiance variation due to the interfering reference and object beams that will be recorded as a thickness variation in the thermoplastic is $(aA \cos \phi)/Z$. This thickness variation is

$$\Delta T = \frac{-\gamma(aA \cos \phi)}{Z} \tag{11-15}$$

Note that the maximum thickness variation $\Delta T_{max} = \gamma aA/Z$. Equation (11-14) can be rewritten as

$$\Delta \psi = k[D \cos \phi] \tag{11-16}$$

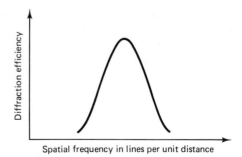

Figure 11-9 Thermoplastic hologram efficiency versus spatial frequency.

where $D = -\gamma(n - 1)aA/Z = -(n - 1)\Delta T_{max}$. The absolute value of D is the maximum variation of the optical path length through the thermoplastic hologram. Equation (11-13) can be rewritten as

$$t = e^{jkD \cos \phi} \tag{11-17}$$

This expression can be expressed in terms of a series of Bessel functions J of order q:

$$e^{jkD \cos \phi} = \sum_{q=-\infty}^{q=\infty} (j)^q \{J_q(kD)\} e^{jq\phi}$$

If the product kD is small,

$$t \simeq jJ_1(kD)e^{-j\phi} + J_0(kD) + jJ_1(kD)e^{j\phi} \tag{11-18}$$

where the first and third terms have the same coefficients because of the fact that

$$J_{-q}(x) = (-1)^q J_q(x)$$

As in the absorption hologram case, we will reconstruct with a playback beam with a complex amplitude of U_P. The product of hologram transmittance and this complex amplitude is

$$U_P t = U_P J_0(kD) + jU_P J_1(kD)e^{j\phi} + jU_P J_1(kD)e^{-j\phi} \tag{11-19}$$

where

$$U_1 = U_P J_0(kD)$$

$$U_3 = jU_P J_1(kD)e^{j\phi}$$

$$U_4 = jU_P J_1(kD)e^{-j\phi}$$

As in the case of the absorption hologram, light transmitted and diffracted by the hologram distributes itself as illustrated in Fig. 11-5, except light described by U_2 is missing. That is, we have assumed that the thermoplastic cannot record low-frequency fringes due to interference of light scattered from the object. The first term U_1 represents light transmitted straight through the hologram. The last two terms represent diffracted light that is capable of producing images. The complex amplitude U_3 describes diffracted light that forms the virtual image due to the phase hologram if the playback beam is equivalent to the reference beam used to produce the hologram. That is, if $U_P = A$,

$$U_3 = jAJ_1(kD)e^{j\phi} \tag{11-20}$$

The first-order Bessel function can be expressed as the series

$$J_1(x) \simeq \frac{x}{2} - \frac{x^2}{2^2 4} + \frac{x^5}{2^2 4^2 6} - \cdots$$

For small values of x, or in this case kD, the first-order Bessel function is approximately $kD/2$. Thus Eq. (11-20) becomes

$$U_3 = jA \frac{kD}{2} e^{j\phi} \qquad (11\text{-}21)$$

Replacing D, this equation becomes

$$U_3 = -j\gamma k(n - 1) \frac{A^2}{2Z} ae^{j\phi} = -[j\gamma k(n - 1)I_R]ae^{j\phi}$$

where I_R is the irradiance of the reconstruction beam. Thus the light represented by U_3 has the same complex amplitude as the original object beam U_0 times j and a multiplicative constant $C = \gamma k(n - 1)I_R$. That is,

$$U_3 = -jCU_0 \qquad (11\text{-}22)$$

Light represented by this term will produce a virtual image of the original object. The diffracted light represented by U_4 can be used to produce a real image. The $-j$ term in Eq. (11-22) indicates that the diffracted light represented by U_3 due to the phase hologram is phase shifted $-\pi/2$ relative to light used to reconstruct the holographic image. This leads to an interesting result in real-time holographic interferometry when using a phase hologram, which will be discussed later.

Figure 11-10 is a plot of Bessel functions of order 0 through 3. It is also interesting to compare the value of $J_0(kD)$ and $J_1(kD)$ and realize values of kD can be chosen for which $J_0(kD)$ is zero. That is, it is theoretically possible to produce a phase hologram for which all the coherent light used for playback will be diffracted, and the transmitted light represented by the term U_1 can be eliminated. However, for larger values of kD, higher-order terms (Bessel functions) must be added to Eq. (11-18). This indicates that

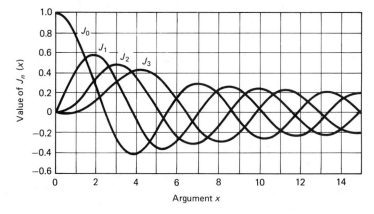

Figure 11-10   Bessel functions of the first kind of order 0 through 3 versus argument x.

for larger values of kD the phase hologram may produce diffracted orders in addition to those represented by U_3 and U_4.

11-2 HOLOGRAPHIC INTERFEROMETRY

Figure 11-11 can be used to obtain the equations needed to determine the phase change of the object beam due to change of its optical path length. Point O is the location of the converging lens focal point, P is a point on the surface of an object, and Q a point on a hologram through which reflected light from the object is observed. The optical path length of the object beam is $r_1 + r_2$. We will consider changes in this optical path length due to displacement of O, P, and index of refraction changes of the media through which the object beam passes. We will first determine the equation for phase change due to the displacement of O and P. The points O' and P' are the displaced positions of O and P. $\mathbf{d}_O$ is the displacement of point O and $\mathbf{d}_P$ the displacement of point P. $\mathbf{k}_1$ and $\mathbf{k}_2$ are propagation constant vectors for the light illuminating P and light scattered to Q, respectively. If the displacements are small, it can be assumed that $\mathbf{r}_1$, $\mathbf{r}_1'$, and $\mathbf{k}_1$ are parallel. The same is true for $\mathbf{r}_2$, $\mathbf{r}_2'$, and $\mathbf{k}_2$. Assuming that the phase of the object beam is zero at point O, the phases of the two beams represented by the optical path lengths $r_1 + r_2$ and $r_1' + r_2'$ at point Q are

$$\phi_1 = \mathbf{k}_1 \cdot \mathbf{r}_1 + \mathbf{k}_2 \cdot \mathbf{r}_2 + \delta \qquad (11\text{-}23)$$

and

$$\phi_2 = \mathbf{k}_1 \cdot \mathbf{r}_1' + \mathbf{k}_2 \cdot \mathbf{r}_2' + \delta \qquad (11\text{-}24)$$

Phase change δ is due to the interaction of the laser light with the object surface. The phase difference of these two optical path lengths is

$$\Delta\phi = \phi_2 - \phi_1 = \mathbf{k}_1 \cdot (\mathbf{r}_1' - \mathbf{r}_1) - \mathbf{k}_2 \cdot (\mathbf{r}_2 - \mathbf{r}_2')$$

From Fig. 11-11 it can be seen that

$$\mathbf{r}_1' - \mathbf{r}_1 = \mathbf{d}_P - \mathbf{d}_O$$

and

$$\mathbf{r}_2 - \mathbf{r}_2' = \mathbf{d}_P$$

Thus

$$\Delta\phi = (\mathbf{k}_1 - \mathbf{k}_2) \cdot \mathbf{d}_P - \mathbf{k}_1 \cdot \mathbf{d}_O \qquad (11\text{-}25)$$

The vector difference $\mathbf{k}_1 - \mathbf{k}_2$ is frequently called the sensitivity vector $\Delta\mathbf{k}$.

Another situation of interest to us is when the optical path length r_2 changes due to the fact the refractive index of the medium between P and Q changes. To illustrate this situation, consider Fig. 11-12, where the re-

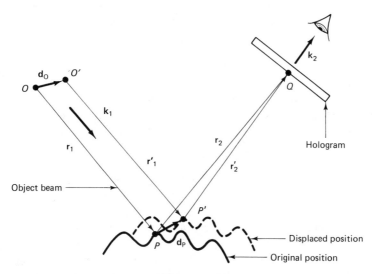

Figure 11-11 Ray diagram for determination of the object beam phase change due to displacement of the object and object beam point source.

fractive index of the medium between the two points is initially 1, and then a medium is injected into the cylindrical region that has a refractive index of n and diameter D. In our illustration we have chosen to eliminate effects due to refraction by investigating a ray that enters and exits the injected medium normal to the interface between the media. The optical path-length

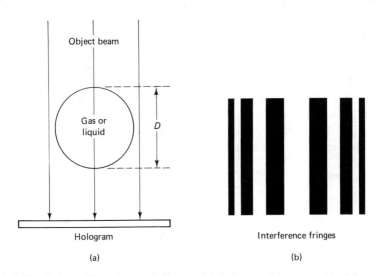

Figure 11-12 Holographic system geometry and interference fringes produced because of change in the optical path length of the object beam due to insertion of a cylindrical region of gas or liquid with refractive index n in an air space.

change due to the injection of the medium will be $(n - 1)\,\Delta r$, and the phase change is

$$\Delta\phi = (n - 1)kD \qquad (11\text{-}26)$$

We can now use Eqs. (11-25) and (11-26) to discuss holographic interferometry. The set of examples used in the following sections are by no means all-inclusive. By studying these examples, students should be able to extend the concepts discussed to other engineering applications of holography.

Double exposure.   The holography system illustrated in Fig. 11-1 can be used for holographic interferometry. In *double exposure holographic interferometry*, the photosensitive medium used to produce a hologram is first exposed with the holography system in some original state. Then a second exposure of photosensitive medium is made after the optical path length of the object beam has been changed due to displacements and/or refractive index changes of the medium through which the object beam travels. The photosensitive medium is developed, producing a hologram after the second exposure. The resulting hologram can be reconstructed using the reference beam or its duplicate. Two images will be reconstructed. Their superposition will produce interference fringes indicative of the phase difference due to the change in optical path lengths occurring between the two exposures. These fringes are similar to topographical map lines superimposed on the object.

In our first example a photographic plate is used as the photosensitive medium. In Problem 9-4, students are asked to prove that the final equations will be the same if a thermoplastic is used to produce a phase hologram. In this example, we wish to determine the displacement magnitude d_P of a center of a circular diaphragm with the holographic system geometry illustrated in Fig. 11-13. The outside perimeter of the diaphragm is clamped, and displacement d_P is known to be in the $+z$ direction. In this example, point O is not displaced and no refractive index changes occur. Under these conditions, Eq. (11-25) reduces to $\Delta\phi = \Delta\mathbf{k}\cdot\mathbf{d}_P$. The vector difference $\Delta\mathbf{k} = \mathbf{k}_1 - \mathbf{k}_2$ is parallel to the bisector of the angle ψ between $\mathbf{k}_1$ and $-\mathbf{k}_2$. Therefore,

$$\Delta\phi = \Delta k\, d_P \cos\phi$$

or

$$\Delta\phi = 2k\, d_P \cos\frac{\psi}{2}\cos\theta \qquad (11\text{-}27)$$

where θ is the angle between the sensitivity vector $\Delta\mathbf{k}$ and the surface displacement $\mathbf{d}_P$. We will assume that the z axis and the displacement are both normal to the diaphragm surface.

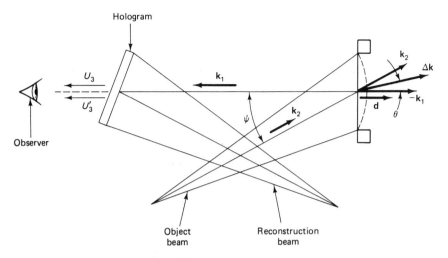

Figure 11-13   System geometry for holographic interferometry.

Recall that the complex amplitude of the light producing the virtual image due to a hologram recorded on a photographic plate is $U_3 = (b\beta I_R)U_O$. To simplify the analysis, we will assume that $b\beta I_R = -1$. The negative sign will be used because the slope of the transmittance versus exposure curve is negative. Thus the virtual image due to the first exposure before the diaphragm center is displaced can be represented by

$$U_3 = -U_O \tag{11-28}$$

and after the displacement by

$$U_3' = -U_O e^{j\Delta\phi} \tag{11-29}$$

where $\Delta\phi$ is the object beam phase change due to displacement d_P. The observer viewing the two images will see fringes due to the interference of U_3 and U_3'. The sum of two terms is

$$U = -U_O(1 + e^{j\phi})$$

and the resulting irradiance is

$$I = 4\left(\frac{U_O U_O^*}{2Z}\right)\cos^2\frac{\Delta\phi}{2}$$

or

$$I = 4I_O \cos^2(\Delta\phi/2) \tag{11-30}$$

where I_O is the irradiance due to either image by itself. The irradiance of the observed light is maximum when

$$\Delta\phi = 2\pi m_B \tag{11-31}$$

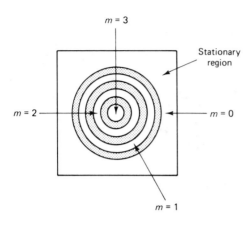

Figure 11-14 Double-exposure holographic image with interference fringes due to normal displacement of the center of a circular diaphragm that is clamped around its perimeter.

where $m_B = 0, 1, 2, 3, \ldots$. Figure 11-14 is a sketch of the resulting interference pattern observed. These interference fringes will appear superimposed on a virtual image of the diaphragm viewed by the observer. By substituting the expression for $\Delta\phi$ from Eq. (11-31) into Eq. (11-27), the surface displacement magnitude d_P is determined to be

$$d_P = \frac{m_B \lambda}{2 \cos (\psi/2) \cos \theta} \tag{11-32}$$

where m_B is an integer representing bright fringes. The stationary region will be represented by $m_B = 0$ and the dark interference fringes by $m_B = \frac{1}{2}, \frac{3}{2}, \frac{5}{2} \ldots$.

Example Problem

Assume that the holographic interferometry configuration illustrated in Fig. 11-13 is used to obtain the double-exposure image sketched in Fig. 11-14. Find the displacement magnitude of the diaphragm center if a HeNe laser is used and $\psi/2 = \theta = 22°$.

Solution The bright fringe at the diaphragm center is represented by $m_B = 3$, and λ for a HeNe laser is 633 ηm. Using Eq. (11-32), we have

$$d_P = \frac{3(633 \text{ ηm})}{2 \cos^2 22°} = 1.10 \text{ μm}$$

In this first example, it was assumed that the displacement direction is known. This often is not the case. When the displacement direction is not known, then the measurement that can be easily determined is the component of the displacement parallel to the sensitivity vector $\Delta\mathbf{k}$, assuming that a stationary portion of the object exists and it can be observed. This component of the displacement is

$$d_{\Delta k} = \frac{m_B \lambda}{2 \cos (\psi/2)} \tag{11-33}$$

It is still not possible, however, to determine if the displacement is in the same or opposite direction of $\Delta \mathbf{k}$. In this text we will discuss the carrier fringe technique to determine displacement direction. Obviously, three components of the displacement can be determined by making three observations from three different angles. Techniques used to determine three-dimensional displacement include using three different holograms or making three different observations through a single hologram. This situation indicates that the fringe location relative to an object changes as an observer views the displaced surface from different angles through the hologram. Such, in fact, is the case. The fringes can exist in a plane other than the object surface. (A more extensive coverage of this concept is contained in Vest, 1979.)

Displacement measurement in many industrial applications is not as important as knowing which portion of an object surface experiences the greatest displacement between two exposures of a hologram. Figure 11-15 is a photograph of a double-exposure hologram image that was produced with a holography system using a double-pulsed ruby laser and reconstructed with a HeNe laser. Here the outside case of a gear box was vibrating at one of its resonant frequencies at a particular operating condition. It is obvious

Figure 11-15   Photograph of reconstructed image from a double-exposure hologram produced using a double-pulsed laser. (Courtesy of Detroit Diesel Allison Division, General Motors Corporation.)

that the surface area located on the right side of the gear box is vibrating with a larger amplitude than in other locations. To reduce vibration and thus the noise produced by the outer surface of this gear box, for the operating condition investigated, this portion of the surface must be treated to reduce motion.

Time average.   Many laboratories do not have a pulsed laser system that can be used for vibration analysis by double-exposure holographic interferometry. If an object vibrates sinusoidally with a constant amplitude, time-average holographic interferometry using a continuous-wave laser provides an alternative.

We will assume that the holography system geometry is the same as that illustrated in Fig. 11-13 and a circular diaphragm is the object to be investigated. As in the example used to explain double-exposure holography, the outside perimeter of the diaphragm is assumed to be clamped, holding it stationary. During the exposure of the hologram, the diaphragm vibrates sinusoidally. The displacement d_P of the diaphragm can be represented by

$$d_P = D \cos \omega_D t \qquad (11\text{-}34)$$

where D is the vibration amplitude and ω_D is the angular frequency of vibration. By using Eq. (11-27), the corresponding phase change of the object beam due to the vibration is

$$\Delta \phi = \alpha \cos \omega_D t \qquad (11\text{-}35)$$

where

$$\alpha = 2kD \cos \left(\frac{\psi}{2}\right) \cos \theta$$

The complex amplitude of the object beam at the hologram can be expressed as

$$U_O(t) = E_0 e^{j\alpha \cos \omega_D t} \qquad (11\text{-}36)$$

If the period of diaphragm vibration is much smaller than the exposure time, information recorded on the hologram is the time average

$$U_O = \frac{1}{T} \int_0^T E_0 e^{j\alpha \cos \omega_D t} \, dt$$

where T is the exposure time. This integration yields

$$U_O = E_0 J_0(\alpha) \qquad (11\text{-}37)$$

where J_0 is a Bessel function of the first kind of order zero.

If a photographic plate is used to produce a time-average hologram, it can be assumed that $U_3 = -U_O$ or $U_3 = -jU_O$ when a phase hologram is

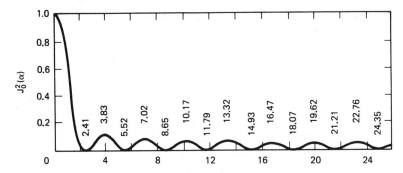

Figure 11-16   Bessel function of the first kind of order zero squared versus argument α.

produced. In either case the irradiance of the light producing the virtual image is

$$I = \frac{E_0^2}{2Z} J_0^2(\alpha) = I_O J_0^2(\alpha) \qquad (11\text{-}38)$$

where I_O is the maximum value of the irradiance at any point on the image.

By using Eq. (11-35), the vibration amplitude D can be determined as a function of α:

$$D = \frac{\alpha\lambda}{4\pi \cos(\psi/2) \cos\theta} \qquad (11\text{-}39)$$

Figure 11-16 is a plot of $J_0^2(\alpha)$ versus α. Note that the positions on the object that do not vibrate will be brighter than other regions. The virtual image of a sinusoidally vibrating circular diaphragm, vibrating at its fundamental resonant frequency, is illustrated in Fig. 11-17. The outside perimeter is very bright, indicating that this region is stationary. The dark fringes correspond to the zero values of the Bessel function. Bright fringes corre-

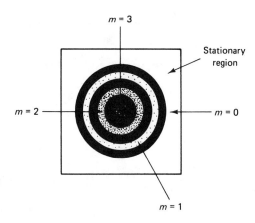

Figure 11-17   Time-average holographic image with interference fringes due to a circular diaphragm vibrating sinusoidally at its fundamental resonant frequency.

spond to the maximum values of $J_0^2(\alpha)$ and have a decreasing irradiance for corresponding larger values of the vibration amplitude D. That is, the center bright fringe on the diaphragm illustrated in Fig. 11-17 has less irradiance than any other bright fringe that represents smaller values of the vibration amplitude.

Example Problem

Determine the vibration amplitude of the circular diaphragm illustrated in Fig. 11-17 if a HeNe laser operating at a wavelength of 633 ηm is used to produce the time-average hologram. Assume that $\theta = \psi = 0$.

Solution   The center bright fringe corresponds to $\alpha = 10.2$. Thus

$$D = \frac{(633 \text{ ηm})(10.2)}{4\pi} = 514 \text{ ηm}$$

Example Problem

Figure 11-18 shows the ray diagram of a hologram system that is used to determine the mode and vibration amplitude of a rod fixed at one end. The rod displacement and direction of the object beam scattered to the photographic plate, used to make the hologram, are both parallel to $+z$. The angle between the object beam illuminating the rod and the z axis is 30°. Determine the vibration mode and vibration amplitude of the rod tip if a HeNe laser operating at 633 ηm is used to produce the hologram.

     Solution   The bright rod base indicates that it is stationary. A vibration node is also indicated by an equally bright area that appears in a region above the base. Therefore, the rod is vibrating in its second harmonic. A dark fringe is located at

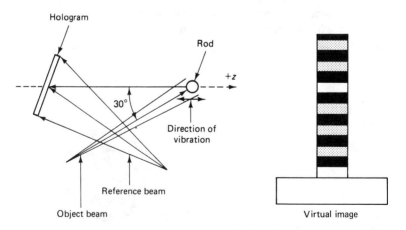

Figure 11-18   Ray diagram of a time-average holography system and virtual image produced with the system to evaluate a vibrating rod.

the rod tip. Using Fig. 11-16, we determine that the dark fringe corresponds to $\alpha = 8.65$. Both $\psi/2$ and θ are equal to $15°$. Thus the vibration amplitude at the rod tip is

$$D = \frac{(8.65)(633 \ \eta m)}{4\pi \cos^2 (15°)} = 467 \ \eta m$$

Real time.   In *real-time holographic interferometry*, the photosensitive medium is exposed, developed, and then placed back in the same position it occupied during exposure. Often this is accomplished by exposing and then developing the photosensitive medium in situ. After the hologram is produced and is located in its original position, the laser is turned back on. The object beam illuminates the object, and the reference beam serves as the playback beam for the hologram. The object is then viewed through the hologram. The virtual image due to the hologram and the object will be located in the same position. If the object experiences a small surface deformation, fringes will be observed due to the interference of the light reflected by the object and the light diffracted by the hologram that forms the virtual image.

Using thermoplastic materials to produce phase holograms has eliminated many of the problems associated with using photographic plates for real-time holographic interferometry. The primary problem is that a photographic emulsion changes thickness during photographic development. This thickness change can shift in the position of the holographic image so that it is not exactly superimposed with the object being evaluated. Thus *background fringes* are produced that may interfere with the information sought. In this section the equations are developed for evaluating the surface displacement of an object using a phase hologram for real-time holographic interferometry.

Let us assume that the geometry of the holography system used to produce a phase hologram is as illustrated in Fig. 11-13. The complex amplitude of the object beam light incident on the recording medium is U_O. After the recording medium is exposed to the interference pattern due to the superposition of the object and reference beam, it is processed to produce a phase hologram. The thermoplastic plate is processed in place. After processing, the object is observed through the phase hologram. Laser light reflected from the object will be observed, as well as the holographic image due to the hologram. To obtain maximum fringe visibility due to object deformation, it is standard practice to equalize the irradiances of the light observed through the hologram that is reflected from the object and the light diffracted by the hologram. This technique adjusts the system so that C in Eq. (11-22) is equal to 1 and $U_3 = -jU_O$. The light reflected from the object can be described by $U'_O = U_O e^{j\Delta\phi}$. The phase shift $\Delta\phi$ is due to the change in the optical path length of the object beam due to object deformation. Since

U_3 and U'_O are superimposed and observed simultaneously, the resultant complex amplitude is

$$U = U_O(-j + e^{j\Delta\phi}) \tag{11-40}$$

The resulting irradiance is

$$I = \frac{U_O U_O^*}{Z} (-j + e^{j\Delta\phi})(j + e^{-j\Delta\phi})$$

or

$$I = I_O[2 + j(e^{j\Delta\phi} - e^{-j\Delta\phi})]$$

where I_O is the object beam irradiance. The expression for the irradiance I simplifies to

$$I = 2I_O(1 - \sin\Delta\phi) \tag{11-41}$$

Irradiance I versus phase shift $\Delta\phi$ is shown in Fig. 11-19. Note that when $\Delta\phi$ is zero the irradiance is $2I_O$, when the phase shift is $-\pi/2$, the irradiance is $4I_O$, and when the phase shift is $\pi/2$, the irradiance is zero. Thus, using a phase hologram, it is possible to find displacement direction of those regions adjacent to the observed stationary regions.

To illustrate our conclusion, assume that the object is a rectangular plate clamped at the bottom and displaced in a direction normal to its surface at the top. The top end of the plate was first displaced toward the hologram and then away from it. The stationary region of the plate has an irradiance of $2I_O$. In Fig. 11-20, this region is shown to be gray. Figure 11-20(a) illustrates the fringes formed when the displacement of the top of the plate is toward the hologram.

Displacement of the plate toward the hologram produces a negative phase shift of the object beam. A bright fringe is produced adjacent to the stationary region of the plate. Displacement of the plate away from the hologram produces a positive phase shift of the object beam, and a dark fringe

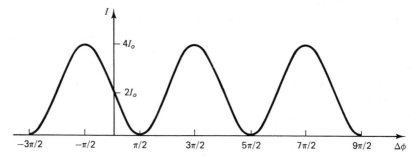

Figure 11-19 Irradiance versus phase shift for a real-time hologram produced using a thermoplastic plate.

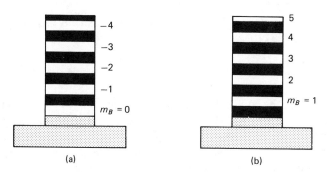

Figure 11-20   Real-time images reconstructed using a thermoplastic plate with interference fringes due to (a) displacement toward and (b) away from the hologram.

is formed adjacent to the stationary region of the plate, as illustrated in Fig. 11-20(b).

To determine the general equation for displacement, bright fringes are assigned an integer number m_B. It can be shown that the irradiance determined by Eq. (11-41) is $4I_O$ when

$$\Delta\phi = \frac{(4m_B - 1)\pi}{2} \qquad (11\text{-}42)$$

where $m_B = \ldots -3, -2, -1, 0, 1, 2, 3 \ldots$ Positive integers are used when a dark fringe is adjacent to the stationary region. Zero and negative integers are used when a bright fringe is adjacent to the stationary region. As in the double-exposure case, the phase shift $\Delta\phi$ is also given by Eq. (11-27). d_P is the displacement of a surface point P on the object, k_1 is the propagation vector for the object beam light illuminating point P, and k_2 is the propagation vector for light reflected from point P onto the phase hologram at the observation point Q. By using Eqs. (11-27) and (11-42) and solving for d_P, we obtain

$$d_P = \frac{(4m_B - 1)\lambda}{8 \cos (\psi/2) \cos \theta} \qquad (11\text{-}43)$$

where ψ is the angle between k_1 and $-k_2$ and θ the angle between the vector difference $k_1 - k_2 = \Delta k$ and the displacement d_P.

In real life a holography system will seldom have the stability required to use the technique described above. In most applications, the observer looking through the hologram at the stationary region of the object can watch the irradiance in this region change over a period of time. Thus, when the holographic image is photographed or videotaped, the stationary region may not have an irradiance of $2I_O$ as indicated. In most cases, a photograph or videotape playback can be chosen for which the stationary region is either bright, $I = 4I_O$, or dark, $I = 0$. This condition can be nicely handled by changing Eq. (11-43) to

$$d_P = \frac{(m_B + m_S)\lambda}{2 \cos (\psi/2) \cos \theta} \tag{11-44}$$

where $m_S = 0$ if the stationary region is bright, $m_S = \frac{1}{2}$ if the stationary region is dark, and $m_B = 0, 1, 2, 3 \ldots$. Equation (11-44) can also be used to determine object displacement at a point relative to any reference point on the object. In this case, m_S is used to represent a dark or bright fringe at the reference point, and m_B is the bright fringe count at the point at which the relative displacement is to be determined.

11-3 CARRIER FRINGES

Most techniques discussed thus far can be used to determine the displacement magnitude of an object but not direction. In 1985, Plotkowski, Hung, Hovanesian, and Gerhart described a carrier fringe technique to determine displacement direction [Plotkowski and others, 1985]. To demonstrate this technique, a rectangular diaphragm clamped around its outside perimeter is our chosen object. A micrometer is used to displace the center of the diaphragm in a direction normal to the plate surface. A schematic of the holography system geometry is as shown in Fig. 11-21. This is a conventional holography system with one modification. The object beam light point source can be displaced. This can be accomplished by reflecting the object beam from a mirror mounted on a rotating stage.

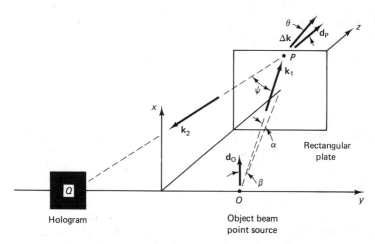

Figure 11-21 A holography interferometry system geometry designed to determine object displacement using carrier fringes.

To facilitate our discussion, the phase shift $\Delta\phi$ due to object displacement is assumed to be $2\pi m_B$, where m_B is an integer representing bright fringes. Using Eq. (11-25),

$$m_B = \frac{(\mathbf{k}_1 - \mathbf{k}_2)\cdot\mathbf{d}_P}{2\pi} - \frac{\mathbf{k}_1\cdot\mathbf{d}_O}{2\pi} \tag{11-45}$$

A real-time hologram is first produced displacing the object center without displacing the object beam point source. That is, $\mathbf{d}_O = 0$. Equation (11-45) then reduces to

$$m_P = \frac{\Delta\mathbf{k}\cdot\mathbf{d}_P}{2\pi} = \frac{2d_P \cos{(\psi/2)} \cos\theta}{\lambda} \tag{11-46}$$

Integer m_P will be called the *displacement fringe number*. A photograph of displacement fringes with bright fringe numbers is shown in Fig. 11-22(a).

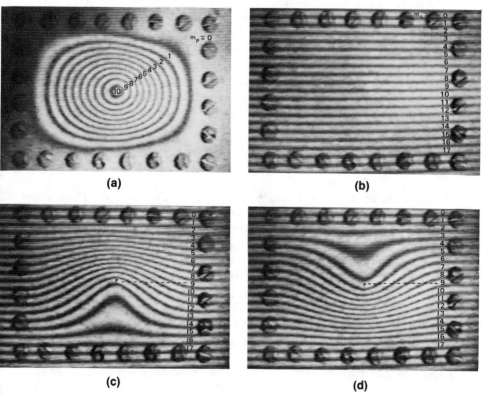

Figure 11-22 Photographs of real-time holographic images with interference fringes due to displacement of a rectangular diaphragm clamped around its perimeter. Interference fringes shown are (a) displacement fringes, (b) carrier fringes due to displacement of the object beam point source in the $+x$ direction, and (c) object displacement in the $+z$ and (d) in the $-z$ direction.

Carrier fringes are then generated by displacing the object beam point source by d_O by rotating the object beam mirror. If the object displacement $d_P = 0$, then Eq. (11-45) can be reduced to

$$m_C = -\frac{k_1 \cdot d_O}{2\pi} = -\frac{d_O \cos \beta}{\lambda} = -\frac{d_O \sin \alpha}{\lambda} \qquad (11\text{-}47)$$

where β is the angle between the two vectors k_1 and d_O, and α is the angle between k_1 and a line perpendicular to d_O. In Fig. 11-21 this line is drawn from object beam point source to the object center. The integer m_C is called the *carrier fringe number*. Figure 11-22(b) is a photograph of carrier fringes generated using real-time holography. If α is limited to small values,

$$m_C \simeq -\frac{d_O \alpha}{\lambda} \qquad (11\text{-}48)$$

Note that m_C is a linear function of α. If d_O is upward ($+x$ direction), it is convenient to assign the uppermost bright fringe a carrier fringe number of zero. The opposite is true when d_O is downward ($-x$ direction). Other bright fringes are assigned positive integer values as illustrated in Fig. 11-22(b).

After carrier fringes are generated, the rectangular plate center is displaced. Figure 11-22(c) is a photograph of the fringes when the rectangular plate center is displaced in the $+z$ direction, and Fig. 11-22(d) in the $-z$ direction. To describe these fringes, Eq. (11-45) can be rewritten $m_B = m_P + m_C$ or

$$d_P = \frac{(m_B - m_C)\lambda}{2 \cos (\psi/2) \cos \theta} \qquad (11\text{-}49)$$

Consider the point P at the center of the plate in Fig. 11-22(c). The carrier fringe number m_C at point P was 9 when $d_P = 0$. This carrier fringe number was determined by constructing a horizontal line from point P over to the stationary region of the object. The bright fringe number m_B is 13. This was determined by following the bright fringe over into the stationary region of the object. In Fig. 11-22(d), the carrier fringe number is 9 and the bright fringe number is 5. Using these numbers, the displacement in Fig. 11-22(c) is determined to be in the $+z$ direction away from the hologram, and in Fig. 11-22(d) it is in the $-z$ direction toward the hologram. In general, if the displacement deforms the carrier fringes in the same direction as d_O, the displacement is away from the hologram, and if the carrier fringes are deformed in the opposite direction of d_O, the displacement is toward the hologram.

Another fringe-counting technique is to observe the carrier fringes at a point on the object and count the number of bright fringes that pass that point as the object is deformed. This count, $\Delta m_P = m_B - m_C$, is positive if the fringes move in the same direction as d_O and negative if they move in the opposite direction. Videotape equipment can be used to obtain a permanent record, and then slow motion or freeze-frame playback is used to obtain fringe counts. The carrier fringe technique can also be used in double-exposure holographic interferometry, but it does not have the advantage of real-time observation of interference fringes as the object is deformed.

Holographic interferometry is being used for nondestructive testing in industry. The time required and the cost of using holographic interferometry can be reduced if computer vision can be used to analyze interference fringes. In some applications, speckle interferometry is a viable alternative to holography.

11-4 ELECTRONIC SPECKLE PATTERN INTERFEROMETRY

Electronic speckle pattern interferometry (ESPI) provides an alternative to holography for analyzing the deformation and vibration of objects. ESPI is a technique by which objects illuminated with coherent light are imaged using lenses, and the data obtained are processed electronically. The primary advantage of this technique is that the hologram is replaced by a two-dimensional detector commonly used in a TV camera, such as a vidicon tube or CCD array. A computer is used to digitize, store, and analyze these images. This technique can detect in-plane or out-of-plane displacements of objects, whereas holography is most effective in measuring out-of-plane displacements. The disadvantages are that a three-dimensional image is not obtained, and the images are formed by lenses and are limited by aberrations and diffraction inherent to these devices.

Speckle.   Coherent light reflected from a diffusive surface produces laser speckle. That is, when viewed the reflected light appears to be made up of bright and dark spots or speckle. This is due to the coherent light reflected from small peaks and valleys on a diffusive surface and interfering on a detector. Figure 11-23 illustrates coherent light being reflected from two points on an object interfering on the image plane of a converging lens. Point source images will have a diameter determined by the aberrations and diffraction due to the lens system. If the lens is diffraction limited, then each point source image is described by the diffraction pattern due to the circular aperture of the lens. If the point sources are close together and their images will overlap and interfere. An object can be thought of as being made up of a large number of point source pairs that have overlaying images that interfere, creating speckle. Experimental results have shown that, for large object distances, bright speckle spot diameter d is given approximately by the equation

$$d \simeq \frac{\lambda F}{D} = \lambda A \qquad (11\text{-}50)$$

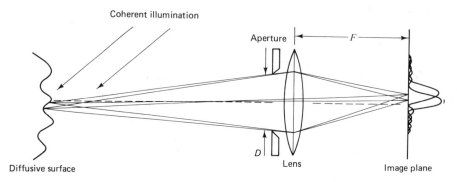

Figure 11-23 Overlapping images of two point sources formed by a diffraction-limited lens.

where λ is the wavelength, F the lens focal length, and D the lens exit pupil diameter. The lens f-number of f/stop is $A = F/D$. Note that speckle size can be conveniently controlled by changing the lens pupil diameter or f/stop. This can be demonstrated by viewing laser speckle through small spaces between your fingers. These small spaces between your fingers decrease the effective aperture diameter of your eye and the speckle will be of larger diameter than when viewed directly.

Digitized images.   Computer vision has provided a technique of digitizing images so they can be stored in a computer. For our purposes we can think of image information as being stored as a set of three-dimensional vectors (x_i, y_i, I_i). The image is divided into equal-sized rectangular picture elements called *pixels*. Pixel arrays of 512×512 are commonly used, and 1024×1024 arrays will be available in the future. Digitized values of x_i and y_i are used to indicate the position of each pixel in the image plane. Image irradiance at each pixel location is digitized as a gray level. Gray levels are now commonly digitized into 256 levels. A level of 0 indicates zero irradiance and 256 indicates the irradiance necessary to saturate the detector. In ESPI we are interested in taking the difference of two digitized images. That is, the gray levels at each pixel site stored from two different images are subtracted and then displayed on a TV monitor. Note that this difference can be either negative or positive. Negative irradiance or a negative gray level, which corresponds to negative light, has no meaning. For this reason, the absolute value of gray level is usually displayed. An alternative would be to set all negative values to zero.

In-plane displacement measurement.   System geometry for a *dual-beam method* of determining in-plane displacement with ESPI is illustrated in Fig. 11-24. Propagation constant vectors $\mathbf{k}_1$ and $\mathbf{k}_2$ indicate the direction of the two coherent beams illuminating point P and its displaced position P' on an object. These vectors are in the same plane, and their angles of in-

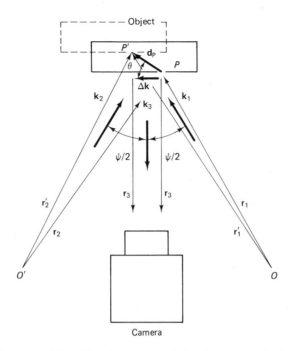

Figure 11-24 Geometry of dual-object beam system designed to measure in-plane displacement of an object using electronic speckle pattern interferometry.

cidence are equal. The direction of the light entering the camera lens after being scattered from point P and P' is indicated by propagation constant vector $\mathbf{k}_3$. Vectors $\mathbf{r}_1$, $\mathbf{r}_2$, $\mathbf{r}_1'$, and $\mathbf{r}_2'$ represent the optical path lengths of the two beams from their respective point sources O and O' to points P and P' on the object. Vector $\mathbf{r}_3$ represents the optical path length from point P or P' to the camera. Assuming that phase is zero at point source O, the phase of light at the camera of the light scattered from point P on the object due to beam 1 is

$$\phi_1 = \mathbf{k}_1\cdot\mathbf{r}_1 + \mathbf{k}_3\cdot\mathbf{r}_3 + \delta$$

Phase change δ is due to the interaction of the laser light with the object surface. Phase of the light at the camera of the light scattered from point P due to beam 2 is

$$\phi_2 = \phi_0 + \mathbf{k}_2\cdot\mathbf{r}_2 + \mathbf{k}_3\cdot\mathbf{r}_3 + \delta$$

Phase ϕ_0 is the light phase at point source O' relative to O. The phase difference of the light at the camera due to the two beams scattered from point P is

$$\Delta\phi = \phi_2 - \phi_1 = \mathbf{k}_2\cdot\mathbf{r}_2 - \mathbf{k}_1\cdot\mathbf{r}_1 + \phi_0 \tag{11-51}$$

The phase difference of the light at the detector due to the two beams scattered from point P' is

$$\Delta\phi + \delta\phi = \mathbf{k}_2 \cdot \mathbf{r}_2' - \mathbf{k}_1 \cdot \mathbf{r}_1' + \phi_0 \tag{11-52}$$

where $\delta\phi$ is the phase shift due to the displacement $\mathbf{d}_P$. Using Eqs. (11-51) and (11-52), we find

$$\delta\phi = \mathbf{k}_2 \cdot (\mathbf{r}_2' - \mathbf{r}_2) - \mathbf{k}_1 \cdot (\mathbf{r}_1' - \mathbf{r}_1)$$

Vector differences $\mathbf{r}_2' - \mathbf{r}_2$ and $\mathbf{r}_1' - \mathbf{r}_1$ are both equal to $\mathbf{d}_P$ thus

$$\delta\phi = (\mathbf{k}_2 - \mathbf{k}_1) \cdot \mathbf{d}_P \tag{11-53}$$

For our system geometry, the propagation constant vector difference $\Delta\mathbf{k} = \mathbf{k}_2 - \mathbf{k}_1$ is parallel to the object surface and

$$\delta\phi = 2kd_P \sin\left(\frac{\psi}{2}\right) \cos\phi \tag{11-54}$$

where ψ is the angle between the propagation constant vectors and θ the angle between the displacement d_P and propagation constant vector difference $\Delta\mathbf{k}$. The displacement component parallel to $\Delta\mathbf{k}$, or in this case parallel to the object surface, is $d_P \cos\theta = d_{\Delta k}$. That is,

$$d_{\Delta k} = \frac{\delta\phi}{2k \sin(\psi/2)} \tag{11-55}$$

Image irradiance of the light scattered from point P is

$$I_P = 2I[1 + \cos\Delta\phi]$$

and from P' is

$$I_{P'} = 2I[1 + \cos(\Delta\phi + \delta\phi)]$$

Image irradiance I has been assumed to be the same for the two beams. Displacement d_P is assumed to be small, and the images of P and P' overlay each other on the detector. The image irradiances I_P and $I_{P'}$ are digitized and stored as gray levels in a computer vision system. The computer is used to take the absolute value of the difference of these two irradiances:

$$|I_{P'} - I_P| = 2I|\cos\Delta\phi - \cos(\Delta\phi + \delta\phi)|$$

or

$$|I_{P'} - I_P| = 4I\left|\sin\left(\Delta\phi + \frac{\delta\phi}{2}\right)\sin\left(\frac{\delta\phi}{2}\right)\right| \tag{11-56}$$

We will evaluate for $|I_{P'} - I_P|$ equal to zero; that is, either sine term or both must be zero. Note that this difference is zero if the irradiance of a speckle spot on the image plane is the same before and after point P is displaced. We will start by analyzing the $\sin(\delta\phi/2)$ term. For dark fringes, $\sin(\delta\phi/2) = 0$ and $\delta\phi = 2m_D\pi$, where $m_D = 0, 1, 2, 3, \ldots$. Using Eq. (11-54), we find

$$d_P = \frac{m_D\lambda}{2\sin(\psi/2)\cos\theta} \tag{11-57}$$

and Eq. (11-55) becomes

$$d_{\Delta k} = \frac{m_D\lambda}{2\sin(\psi/2)} \tag{11-58}$$

Figure 11-25 is a photograph of speckle interference fringes due to in-plane displacement.

Analysis of the $\sin(\Delta\phi + \delta\phi/2)$ in Eq. (11-56) yields even more information about this speckle interferometry technique. For this term to be zero, $(\Delta\phi_1 + \delta\phi/2) = n_D\pi$. It is interesting to investigate the term in the bright fringe regions, that is, $\delta\phi = (2m_B + 1)\pi$. Students will discover by working Problem 11-18 that conditions exist for which the image irradiance at points in the bright fringe regions is zero (see Fig. 11-25). For this reason, speckle fringes obtained by this ESPI technique do not have as much contrast as the fringes obtained by using holography.

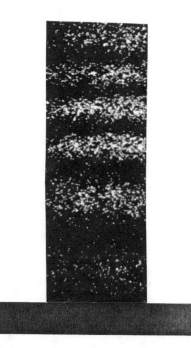

Figure 11-25 Speckle interferometry fringes due to in-plane displacement of a cantilever beam.

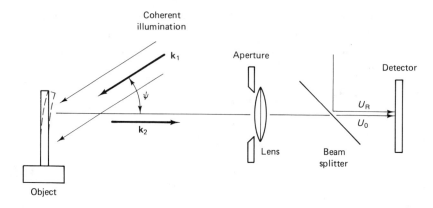

Figure 11-26 Geometry of a reference-beam system designed to measure out-of-plane displacement of an object using electronic speckle pattern interferometry.

Out-of-plane displacement measurement. The *reference-beam method* will be used to determine out-of-plane displacement. Equations that can be used to determine out-of-plane displacement using ESPI are similar to those used for double-exposure holographic interferometry. Figure 11-26 is an illustration of a system geometry that can be used. The object is illuminated with coherent light and is imaged by a TV or camera lens with its corresponding speckle pattern on a two-dimensional detector. This image is superimposed with a reference beam. If the complex amplitude of the reference beam is assumed to be $U_R = A$ and the object beam $U_O = ae^{j\phi}$, before the displacement of point P, the resulting irradiance I_1 due to this superposition is described by Eq. (11-12):

$$I_1 = I_R + I_O + 2\sqrt{I_R I_O} \cos \phi$$

Phase difference between the object and reference beam is ϕ. The first image is digitized and stored. After deformation, the resulting image irradiance of the displaced point P' is

$$I_2 = I_R + I_O + 2\sqrt{I_R I_O} \cos (\phi + \delta\phi)$$

where the phase shift $\delta\phi$ is due to the displacement $\mathbf{d}_P$. The second digitized image is stored and subtracted. If the absolute value of the difference is used,

$$|I_2 - I_1| = 2\sqrt{I_R I_O} \, |\cos \phi - \cos (\phi + \delta\phi)|$$

or

$$|I_2 - I_1| = 4\sqrt{I_R I_O} \, |\sin(\phi + \delta\phi/2) \sin(\delta\phi/2)| \qquad (11-59)$$

As in the development of the equations for using ESPI to determine in-plane displacements, dark fringes accrue when $\delta\phi = 2m_D\pi$. Surface displacement $\mathbf{d}_P$ is related to the phase shift $\delta\phi$ by Eq. (11-27). If $\mathbf{d}_O$ is zero, this equation reduces to

$$d_P = \frac{m_D\lambda}{2\cos(\psi/2)\cos\theta} \tag{11-60}$$

The only difference between this equation and Eq. (11-32), derived for double-exposure holographic interferometry, is that the integer m_D represents dark fringes rather than bright fringes. Note that Eq. (11-27) indicates that carrier fringes can also be generated to determine displacement direction. Again, the contrast of the fringes generated by ESPI is not expected to be as good as those generated by holography because of the term $\sin(\phi + \delta\phi/2)$ contained in Eq. (11-59).

The contrast of fringes ESPI can be improved by using image enhancement techniques developed for computer vision. Computer vision systems with sufficient processing speeds can be used to analyze vibrating objects. By digitizing images at more than two positions during deformation or vibration, static and dynamic high-quality holographic interferometry fringes can be obtained. Figure 11-27 is a photograph of time-average fringes obtained by this technique.

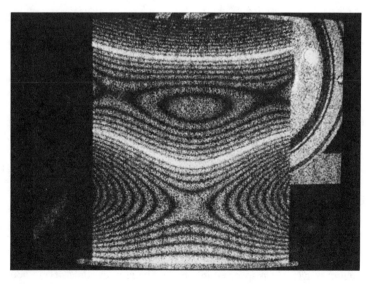

Figure 11-27 Photograph of dynamic holographic interferometry fringes. (Courtesy of United Technologies.)

PROBLEMS

11-1. Describe the results if a hologram is reconstructed with the object beam.

11-2. The system geometries used to record transmission and reflection holograms are illustrated in Fig. 11-28. The reference beam U_R is collimated and object beam U_O originates from a point source. By using Eq. (11-7) and drawing sketches, describe the results when the transmission and reflection holograms are reconstructed with:

(a) $U_P = U_R$
(b) $U_P = U_R^*$
(c) $U_P = U_O$
(d) $U_P = U_O^*$

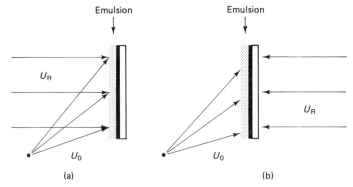

Figure 11-28 Recording geometries for (a) transmission and (b) reflection holograms.

State which playback schemes produce real and virtual images of the point source.

11-3. Assume that a HeNe laser and a thermoplastic plate with a 1.6 refractive index are used to produce a phase hologram.

(a) Find the amplitude of the thickness variation Δt_{max} of the thermoplastic plate required for $J_0(kD)$ to equal $J_1(kD)$. Explain the consequence of this equality.

(b) Find the amplitude of the thickness variation Δt_{max} required for $J_0(kD)$ to be zero. For this value, find the value of $J_1(kD)$ and $J_2(kD)$. Explain the consequence of this condition.

11-4. Assume that the complex amplitude of the light producing the virtual image of a hologram recorded on thermoplastic is given by

$$U_3 = -jU_O$$

In the case of double-exposure holographic interferometry, show that Eq. (11-30) describes image irradiance I as a function of phase shift $\Delta \phi$.

11-5. The system geometry illustrated in Fig. 11-12 is used to analyze a cylindrical region of gas that is inserted into an air space in the time period between the exposures used to produce a double-exposure hologram. If a HeNe laser is used and the cylindrical region has a diameter of 4 cm, determine the refractive index of the gas.

11-6. Using the holography system and virtual image illustrated in Fig. 11-18, determine the order and type of fringe located at the tip of the vibrating rod if a double-pulsed ruby laser is used to produce a double-exposure hologram, with the two exposures taken with the rod in positions of maximum positive and negative displacement relative to equilibrium.

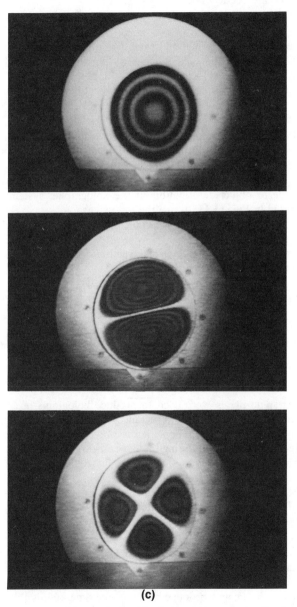

(c)

Figure 11-29 Time-average fringes obtained by holographic analysis of a circular diaphragm vibrating at three different resonant frequencies.

11-7. Figure 11-29 is a collection of photographs of time-average hologram images of a diaphragm vibrating at different resonant frequencies. A HeNe laser was used to produce the holograms. The propagation vectors k_1 and k_2 make an angle of 30° and 0° relative to the diaphragm surface normal, respectively.
  (a) With a written statement, describe how the diaphragm is vibrating at the resonant frequencies.
  (b) Determine the largest vibration amplitude for the highest-order resonant mode of vibration.

11-8. Assume that the deformation of the diaphragm shown in Fig. 11-22(a) is to be analyzed using the system geometry illustrated in Fig. 11-21. The tenth-order bright fringe is located at P (20, 10, 50), the viewing position through the hologram at Q (0, −20, 0), and the object beam point source at O (0, 20, 0). If a HeNe laser is used and all distances are measured in centimeters, determine the displacement component parallel to the z axis at point P.

11-9. Assume that a photographic plate is used as the recording medium in a real-time holographic interferometry experiment and that the virtual image is described by $U_3 = -U_O$. Also, the light reflected from the object and viewed through the hologram is described by $U'_O = U_O e^{j\Delta\phi}$, where $\Delta\phi$ is the phase shift due to object deformation. If the holographic system geometry is as illustrated in Fig. 11-13, derive the following:
  (a) The equation for irradiance due to the superposition of U_3 and U'_O as a function $\Delta\phi$.
  (b) The equation for surface displacement d_P as a function of an integral number representing fringes.

11-10. An argon laser operating at 514 ηm and a thermoplastic holographic plate are used in a real-time holographic interferometry experiment. The system geometry is illustrated in Fig. 11-18. The fringes observed through the hologram are as illustrated in Fig. 11-20. Assume the displacement of the rectangular plate is parallel to the z axis. Determine the displacement magnitude and direction of the top of the rectangular plate for:
  (a) Fig. 11-20(a).
  (b) Fig. 11-20(b).

11-11. Photographic plates were used to produce holograms in an interferometry experiment. The object was a rectangular steel plate clamped at one end as illustrated in Fig. 11-18. The series of photographs shown in Fig. 11-30 are of the images observed during the experiment.
  (a) State which is a photograph of the image due to a real-time, double-exposure, or time-average hologram.
  (b) State the reasons for your conclusions.
  (c) State the similarities and differences in these images if a thermoplastic holographic plate were used to record the holograms.

11-12. The system geometry used in the experiment described in Problem 11-11 is identical to the one illustrated in Fig. 11-18. A HeNe laser operating at 633 ηm and photographic plates were used to produce the holograms.
  (a) Determine the displacement magnitude of the top of the steel plate by using the images produced by the real-time and double-exposure holograms.

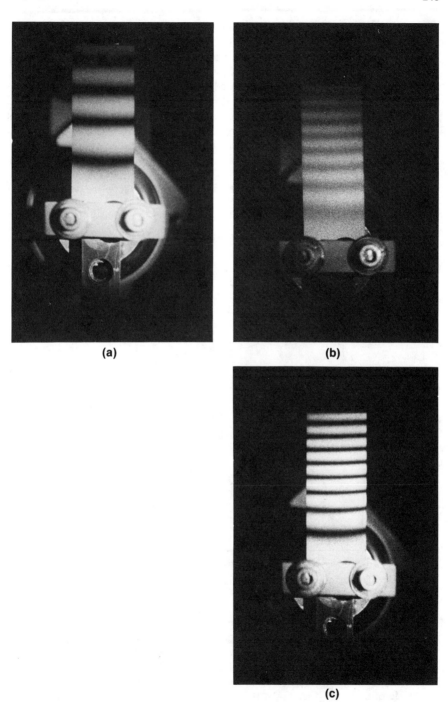

(a)

(b)

(c)

Figure 11-30 Photographs of interference fringes obtained using three different types of holographic interferometry.

(b) Determine the vibration amplitude of the top of the steel plate by using the image produced by the time-average hologram.

11-13. Figure 11-31 is a tracing of dark fringes from a photograph obtained from a real-time holographic interferometry experiment. The object was a steel panel held stationary along its bottom edge and displaced in the $+z$ direction at the top edge. The system geometry was as illustrated in Fig. 11-21, except the angles ψ and θ were small. An argon laser operating at 514 ηm and thermoplastic plates were used to produce the holograms. The image of the stationary edge appeared bright when observed through the hologram.

(a) Using small angle approximations and the proper value of m_s, obtain a simplified version of Eq. (11-44) that can be used in this application.

(b) Plot the normal displacement magnitude of the steel plate versus position along the x axis.

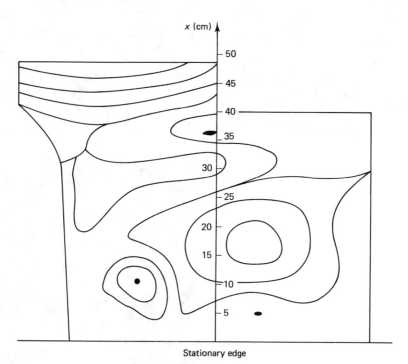

Figure 11-31 Tracing of dark interference fringes due to displacement of a sheet-metal panel obtained by using real-time holographic interferometry.

11-14. A panel similar to the one discussed in Problem 11-12 was evaluated using the carrier fringe technique. An argon laser operating at 514 ηm and thermoplastic plates were used to produce the holograms. The system geometry used was as illustrated in Fig. 11-21, except the angles ψ and θ were small. Figure 11-32(a) is a tracing of the dark carrier fringes before the panel was deformed, and Fig. 11-32(b) is after deformation. The displacement of the object beam focal point $\mathbf{d}_O$ was in the $-x$ direction.

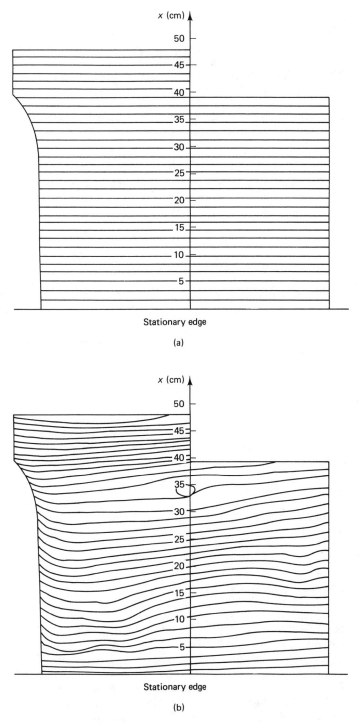

Figure 11-32 Tracing of dark carrier fringes (a) before and (b) after deformation of a sheet-metal panel obtained using real-time holographic interferometry.

(a) Using small angle approximations, obtain a simplified version of Eq. (11-49) that can be used in this application.

(b) Plot the normal displacement magnitude of the steel plate versus position along the x axis.

11-15. In a holographic interferometry experiment, the light producing the image due to a real-time hologram can be described by

$$U_3 = U_O e^{j\Delta\phi_R}$$

where $\Delta\phi_R$ is a phase change due to a change of the optical path length of the reference beam. The laser light reflected from the object and viewed through the hologram is described by

$$U'_O = U_O e^{j\Delta\phi}$$

where the phase change is due to a displacement $\mathbf{d}_P$.

(a) Derive the equation for the irradiance due to the interference of U_3 and U'_O as a function of $\Delta\phi_R$ and $\Delta\phi$.

(b) Derive the equation for the integer m_B, which represents bright fringes, in terms of k, d_P, ψ, θ, $\Delta\phi$, and $\Delta\phi_R$.

(c) If the system geometry is as illustrated in Fig. 11-21 and $\mathbf{d}_P$ is in the $+z$ direction, determine the effect on m_B if the phase change $\Delta\phi_R$ increases. What is the effect on m_B if $\mathbf{d}_P$ is in the $-z$ direction and $\Delta\phi_R$ increases? If Fig. 11-22(a) is the image observed through the real-time hologram, how could this concept be used to determine displacement direction?

11-16. Assume that an argon laser operating at 514 ηm and the ESPI configuration illustrated in Fig. 11-24 are used to analyze the cantilever beam shown in Fig. 11-25. For $\psi/2 = 15°$, determine the displacement component parallel to $\Delta\mathbf{k}$ at the top of the beam.

11-17. Assume that the speckle fringes shown in Fig. 11-25 are obtained using the ESPI system illustrated in Fig. 11-26 and an argon laser operating at 514 ηm. If $\psi = 30°$ and the beam displacement is parallel to $\mathbf{k}_2$, determine the displacement at the top of the beam.

11-18. By using Eq. 11-59, derive equations for ϕ in terms of a set of integers that describes the conditions for bright and dark speckle spots in the bright fringe regions where $\sin(\delta\phi/2) = \pm 1$.

11-19. Assume that an argon laser operating at 488 ηm and the configuration illustrated in Fig. 11-21 are used to obtain the time-average holographic fringes shown in Fig. 11-27 using a reference-beam multimage ESPI technique. Assume that the middle of the bottom edge of the plate is located at $(0, 0, 50)$, the camera at point Q $(0, 0, 0)$, and the object beam point source O at $(0, 20, 0)$. All measurements are in centimeters. Assume the plate is 20 cm wide and 25 cm tall. Using Eq. (11-39) or its equivalent, plot the vibration amplitude as a function of position along a vertical line (a) through the middle of the plate and (b) along the right side of the plate.

REFERENCES

CAULFIELD, H. J. (ed.), *Handbook of Optical Holography*. New York: Academic Press, Inc., 1979.

COLLIER, R. J., C. B. BURCKHARD, AND L. H. LIN, *Optical Holography*. New York: Academic Press, Inc., 1971.

ERF, R. K. (ed.), *Speckle Metrology*. New York: Academic Press, Inc., 1978.

GASKILL, J. D., *Linear Systems, Fourier Transforms, and Optics*. New York: John Wiley & Sons, Inc., 1978.

GOODMAN, J. W., *Introduction to Fourier Optics*. New York: Academic Press, Inc., 1968.

LUXON, J. T., D. E. PARKER, AND P. D. PLOTKOWSKI, *Lasers in Manufacturing*. New York: Springer-Verlag, 1987.

PLOTKOWSKI, P. D., Y. Y. HUNG, AND J. D. HOVANESIAN, "An Improved Fringe Carrier Technique for Unambiguous Determination of Holographically Recorded Displacements," *Optical Engineering*, **24**, October 1985, 754–56.

VEST, C. M., *Holographic Interferometry*. New York: John Wiley & Sons, Inc., 1979.

Chapter 12

Interaction
of High-Power Laser Beams
with Materials

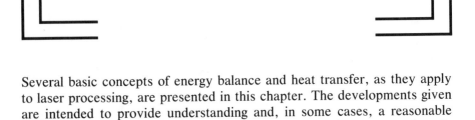

Several basic concepts of energy balance and heat transfer, as they apply to laser processing, are presented in this chapter. The developments given are intended to provide understanding and, in some cases, a reasonable approximation (at least order of magnitude) to reality. The types of applications to which the concepts developed here are applicable are heat treatment, welding, and material removal. Alloying, cladding, and glazing, less important industrially at this time, can also be accomplished.

The interaction of a laser beam with the workpiece depends on several laser beam parameters as well as material parameters. These parameters are discussed prior to a treatment of energy balance and heat-transfer concepts.

12-1 MATERIAL AND LASER PARAMETERS

Material and laser parameters that affect laser processing are discussed in this section.

Material parameters

1. Reflectance R is the ratio of the power reflected from a surface to the power incident on it. In Chapter 1 it was pointed out that R, for normal incidence, is related to the refractive index by

$$R = \left(\frac{n_1 - n_2}{n_1 + n_2}\right)^2 \qquad (12\text{-}1)$$

where n_2 and n_1 are the refractive indices of the substrate material and incident medium, respectively. For conducting or absorbing dielectric materials, n_2 is a complex number and the square in Eq. (12-1) is an absolute square. The magnitude of the refractive index for good conductors (metals) is proportional to $\sqrt{\sigma/2\pi\mu f}$, where σ is electrical conductivity, μ is magnetic permeability, and f is the frequency of the light. Consequently, metals like copper and silver have high reflectances that increase with decreasing frequency (increasing wavelength). It is found that the reflectance of metals substantially decreases as the temperature nears the melting point.

2. Absorption coefficient α is the fractional loss of light power per unit distance for light traveling in a nonmetallic material. Beer's (Lambert's) law relates power to absorption by

$$P = P_0 e^{-\alpha z} \qquad (12\text{-}2)$$

where P_0 is the power entering the surface and P is the power at depth z. The absorption coefficient can also be interpreted as the penetration depth or distance at which the power has dropped to one over e (37%) of the value entering the surface.

3. Specific heat C is the energy required to raise the temperature of unit mass one degree. The SI units are joules per kilogram-Celsius degrees, but calories per gram-Celsius and BTU (British thermal unit) per pound-Fahrenheit degrees are also common. A frequently useful variation is the volume specific heat C_v given by ρC, where ρ is the mass density of the material. C_v is the energy required to raise the temperature of unit volume one degree, or energy per unit volume-degree.

4. Thermal conductivity k is the heat flow per unit area per unit thermal gradient. The units are presented in a variety of ways, but a look at the one-dimensional heat conduction equation, which relates rate of heat flow Q to thermal gradient dT/dz,

$$Q = -kA \frac{dT}{dz} \qquad (12\text{-}3)$$

indicates that the SI units for k are W/m·°C, but W/cm·°C are frequently used.

5. Thermal diffusivity κ is related to thermal conductivity and volume specific heat by

$$\kappa = \frac{k}{C_v} \qquad (12\text{-}4)$$

and is a measure of how much temperature rise will be caused by a pulse of heat applied to the material. It also indicates how rapidly heat will diffuse through the material. Units are generally cm^2/s. Materials with high thermal diffusivity will experience a relatively small temperature rise, with good heat penetration for a given heat pulse at the surface. A material with a low thermal diffusivity will undergo a relatively large temperature rise at the surface, with low heat penetration into the material for a given heat pulse.

6. Latent heat L refers to the amount of heat required to cause a change of phase of unit mass of material. L_f is the latent heat of fusion or the energy required to cause melting of unit mass. Examples of units are cal/g, BTU/lb, and the SI units J/kg. L_v is the latent heat of vaporization and has the same meaning and units as L_f. Latent heats of vaporization are much larger than latent heats of fusion, and vaporization also requires more energy than that needed to raise unit mass up to the vaporization temperature. Consequently, when vaporization is involved in processing, it is a major factor in determining energy requirements.

7. Transformation temperatures refer to the melting temperature T_m, vaporization temperature T_v, and other phase change temperatures T_p, such as the martensitic transformation temperature on the iron–carbon phase diagram. The last is particularly important in heat treatment of cast iron and steel.

Laser beam parameters.   Laser parameters relate to both the laser beam and to the laser power output as a function of time. Normally, lasers are either operated CW or pulsed; the pulses may be produced by Q-switching. The parameters of the beam that affect processing are as follows:

1. Wavelength
2. Focused spot size
3. Mode structure

Wavelength affects absorption and reflection characteristics, and spot size and mode structure affect average irradiance and irradiance distribution in the spot.

The power output and traverse speed, or dwell time, are clearly important for CW operation. In pulsed lasers the length of the pulse, energy per pulse, peak power, and even the shape of the pulse influence the interaction of the beam and the workpiece.

Pulses from solid lasers usually consist of a substructure or spiking. The details of this spiking may have a significant influence on the quality of drilled holes.

12-2 BASIC HEAT-TRANSFER EQUATIONS

The basic heat-transfer equations considered are

$$\mathbf{F} = -k\,\nabla T \tag{12-5}$$

which relates heat flux $\mathbf{F}$ to thermal gradient T and

$$\nabla^2 T - \frac{1}{\kappa}\frac{\partial T}{\partial t} = \frac{-A}{k} \tag{12-6}$$

which applies to nonsteady-state heat conduction, with A accounting for internal generation of heat with units of W/m^3 or equivalent. A is zero in materials where the incident laser power is absorbed at the surface. This statement will be assumed for all remaining sections of this chapter. When the laser beam penetrates the material to any extent, the solution of Eq. (12-6) requires consideration of a distribution of internal sources and $A(z, t)$ cannot be neglected.

   Equation (12-5) is sufficient for steady-state heat-transfer problems. Unfortunately, laser applications seldom lead to steady-state situations. For the purposes of this book, detailed solutions of Eq. (12-6) will not be undertaken, although solutions that pertain to laser applications will be presented and discussed. Generally, radiation and convection are not significant sources of heat loss in laser processing and will be assumed negligible.

12-3 UNIFORM, CONSTANT IRRADIANCE MODEL

In this section the complete solution for the uniform, constant irradiance model is given. This model gives good results in heat treating applications and results of this model are used to improve the energy balance approach given in section 12-4.

   In this model it is assumed that a semi-infinite solid is irradiated with a constant and uniform power per unit area. In all equations the irradiance I_0 at the surface is assumed to be the actual power per unit area absorbed, $(1 - R)I_i$, where I_i is the incident irradiance. The solution of Eq. (12-6) for the temperature, $T(z, t)$, as a function of depth z into the solid and time t is [Carslaw and Jaeger, 1959]

$$T(z, t) = \frac{2I_0}{k}\left[\left(\frac{\kappa t}{\pi}\right)^{1/2} e^{(-z^2/4\kappa t)} - \frac{z}{2}\,\mathrm{erfc}\,\frac{z}{\sqrt{4\kappa t}}\right] \tag{12-7}$$

where

$$\mathrm{erf}(s) = \frac{2}{\sqrt{\pi}}\int_0^s e^{-x^2}\,dx$$

is the error function and erfc(s) is the complementary error function or $1 -$ erf(s). Tables of the error function are available in most compilations of mathematical tables, such as Jahnke and Emde (1945). The surface temperature of the semi-infinite solid is given by

$$T_s(0,\ t) = \frac{I_0}{k}\left(\frac{4\kappa t}{\pi}\right)^{1/2} \tag{12-8}$$

The graph in Fig. 12-1 was generated from Eqs. (12-7) and (12-8). Equation (12-8) is useful for estimating the time required to reach a specified surface temperature for a given irradiance.

In the case of a pulse of length t_p, $I(t) = I_0$, $T(z,\ t)$ is given by Eq. (12-7) for the duration of the pulse. For $t > t_p$, a solution is obtained by the principle of superposition; thus for $t > t_p$

$$T(z,\ t) = [f(t) - f(t - t_p)] \tag{12-9}$$

where $f(t)$ is used to represent the right side of Eq. (12-8).

Equations (12-7) and (12-9) are valid when the thickness of the part exceeds $\sqrt{4\kappa t}$ and will give a reasonable estimate of the temperature along the axis of the beam extending into the part if the beam width exceeds $\sqrt{4\kappa t}$. Equations (12-7) and (12-9) are most accurately applied to laser heat treatment when a defocused beam is used, possibly in conjunction with beam scanning (dithering), rotation, or beam integration, to spread the power out in a more or less uniform fashion over a fairly large area.

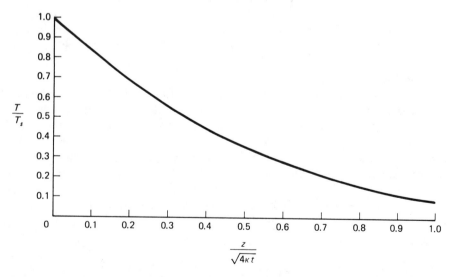

Figure 12-1 Ratio of temperature T at depth z to the surface temperature T, versus $z/\sqrt{4\kappa\tau}$ for uniform, constant irradiance.

12-4 ENERGY BALANCE APPROXIMATION

Simple energy balance approximations frequently provide reasonable ball-park numbers for many applications. For that reason, some discussion of this approach is presented here.

Figure 12-2 is a schematic representation of a laser beam focused onto the surface of a workpiece. If it is assumed that the material is heated to a depth z with cross-sectional area πa^2, then the energy U required to bring the material to temperature T is given by

$$U = (CT + L_f + L_v)\rho \pi a^2 z \qquad (12\text{-}10)$$

where T is used for temperature to avoid confusion with time t and represents the change in temperature of the part. It has been assumed in Eq. (12-10) that ρ and C_v are independent of temperature and are the same for liquid and solid. Equation (12-10) includes the possibility of melting and vaporization. Only the first term is required for surface hardening without melting; the first two terms are used for welding; and all three are applicable for material removal, although the first two may be negligible. The depth of the cylinder heated in Fig. 12-1 may be taken as the thickness of the part, if complete penetration is desired, or to whatever depth requires heating. The depth is typically 0.25 to 1.5 mm in surface hardening.

Obviously, the energy balance approach is crude; it ignores lateral heat transfer and assumes uniform heating throughout the cylinder. Nevertheless, it provides a lower limit to the energy required in a given case and is fairly accurate for small z or $z < a$ and for t relatively short compared with the time required for significant heat loss by conduction.

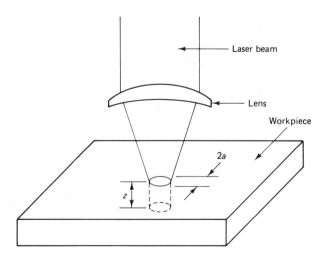

Figure 12-2 Laser beam focused on a workpiece.

Once U is determined, the length of the pulse, assuming constant power P, is just $t = U/P$. The solution of Eq. (12-6) for constant uniform irradiance on the surface of a semi-infinite solid leads to a concept of thermal penetration depth, z_{th}, where

$$z_{th} = \sqrt{4\kappa t} \qquad (12\text{-}11)$$

Equation (12-11) can be used to estimate z for a given pulse length or, conversely, it can be used to estimate the pulse length required to achieve a desired penetration depth. Actually, the temperature at $z = \sqrt{4\kappa t}$ is approximately one-tenth the temperature at the surface, relative to the initial temperature. Figure 12-1 is a plot of $T(z)/T_s$ against $\alpha = z/\sqrt{4\kappa t}$, where T_s is the temperature at the surface of the part. A depth of $z = \sqrt{\kappa t}$ is frequently used to estimate thermal penetration depth and is perhaps more reasonable, for at that depth the temperature change is about 35% of the surface temperature change.

The energy balance approach can be used with a moving source. Assume that the source in Fig. 12-2 is moving to the right with speed v (or the part is moving to the left with speed $-v$). The time that the beam remains on any one point is $t = 2a/v$. Then the energy delivered to a spot of radius a is $U = Pt$ or

$$U = \frac{2aP}{v} \qquad (12\text{-}12)$$

Equation (12-12) can be used in conjunction with Eqs. (12-9) and (12-10) for problems involving a moving source or part. This treatment is quite good for nonmetals with high absorption coefficients.

12-5 HEATING WITH MELTING

The process of melting a material with the energy supplied by the laser is relatively simple to analyze if the time to bring the surface to the melting point is short compared with the thermal time constant. The time to reach melting at the surface t_m can be estimated from the uniform irradiance model by

$$t_m = \frac{\pi}{4\kappa} \left(\frac{kT_m}{I_0} \right)^2 \qquad (12\text{-}13)$$

This is usually much less than a microsecond for typical welding situations. On the basis of energy balance, the volume of material melted can be estimated as

$$V = \frac{U}{\rho(CT_m + L_v)} \qquad (12\text{-}14)$$

where U is the total energy input. Because conduction losses have been neglected, U is a lower limit to the energy required. Doubling U usually gives a reasonable estimate of the actual energy needed. The length of time to reach vaporization can be estimated from

$$t_v = \frac{\pi}{4\kappa} \left(\frac{kT_v}{I_0} \right)^2 \tag{12-15}$$

It is typically milliseconds for pulsed welding with I_0 on the order of 10^5 W/cm^2. The parameters κ and k are not constant with temperature, but reasonable calculations can be made by selecting an intermediate value for the range of temperatures under consideration. Because more time is spent in the molten state than the solid state during heating, the most appropriate values of κ and k are averages over their values in the molten range of interest.

As an example, consider the welding of steel with a CO_2 laser capable of providing a power of 2.5 kW at the workpiece. A weld depth of 2 mm is desired with a weld width of about 1 mm. The appropriate parameters are

$$T_m = 1547°C, \qquad T_v = 2752°C$$

Room temperature is $T_a = 16°C$, $\rho = 7.87$ g/cm^3, $L_f = 272$ J/g, $\kappa = 0.21$ cm^2/s, $k = 0.75$ W/cm °C, and $C = 0.46$ J/g °C.

According to energy balance, $U = \rho V[CT_m + L_f] = 11$ J. U should be doubled to compensate for heat conduction losses. Assuming that half the power is coupled into the part, $t_p = U/P = 19$ ms. In actual practice, a high-order CO_2 laser beam focused to a spot diameter of approximately 0.5 mm was used with a traverse speed of 2.54 cm/s. The dwell time of this beam on any one spot is $t = 0.05$ cm/(2.54 cm/s) = 20 ms. The close agreement of these times is coincidental, for the choice of a 1.0-mm wide weld and 50% reflection are somewhat arbitrary. According to the uniform irradiance model, the time to reach T_v at the surface is

$$t_v = \frac{\pi}{4\kappa} \left(\frac{kT_v}{I} \right)^2 = 0.04 \text{ ms}$$

Obviously, this time is much too short to allow any useful penetration without substantial material removal occurring. The development of a keyhole[1] in the high-power continuous welding, which allows the beam to penetrate unimpeded deep into the part, is absolutely essential and is what makes possible a reasonable estimate using the energy balance approach. It is believed, however, that nearly all the incident energy is coupled into the part during keyholing. An irradiance of 5×10^5 W/cm^2 (the irradiance used in this example) will tend to drill except in the case of continuous welding where a keyhole develops.

---

[1] The phenomenon of keyholing is discussed in more detail in Section 12-8.

The uniform irradiance solution gives reasonable results in micro-welding of small, thin parts. If the penetration depth l and I_0 are known, the graph in Fig. 12-1 can be used to find a value of α at $T(z)/T_s = T_m/t_v$. Then the pulse length is obtained from $\alpha = l/\sqrt{4\kappa t_p}$.

12-6 MATERIAL REMOVAL

A simplified empirical approach to material removal in nonmetals is based on measured values of specific energy Q. Specific energy is an expression that is used to describe either the energy per unit volume or per unit mass required to vaporize the material. Here it will be used to represent energy per unit volume. From Eq. (12-10), the specific energy is given by

$$Q = \frac{U}{V} = (CT_m + L_f + L_v) \tag{12-16}$$

Thus the specific energy can, in principle, be determined theoretically. Since there is little heat conduction loss during drilling or cutting of plastics and ceramics, this approach is fairly reliable.

To drill a hole of depth d, the energy required is $U = VQ$, where V is the volume of the hole. The length of the pulse and laser power may be adjusted to give the appropriate energy. For cutting applications, the power required is given by

$$P = \frac{\pi a Q\ dv}{2} \tag{12-17}$$

where $U = Pt$ and $t = 2a/v$. Table 12-1 contains a list of specific energies for several common materials.

The numbers given in Table 12-1 are representative values only. Actual values of specific energy for plywood, for example, range from 2 to 10 kJ/cm^3 depending on the type of wood and bonding agent and whether it is cut with or across the grain.

TABLE 12-1   Specific Energies in kJ/cm^3

Plywood	7.9
Plexiglass	7.9
Glass	78.0
Concrete	42.0
Boron epoxy	69.0
Fiber-glass epoxy	36.0

12-7 HEATING WITH VAPORIZATION

At power densities sufficient to produce rapid vaporization, typically 10^6 W/cm^2 and higher, the time to reach vaporization can be estimated from Eq. (12-7) because very little heat penetration occurs in this length of time. Using the data for steel and an irradiance of 10^6 W/cm^2, for example,

$$t_v = \frac{\pi}{4\kappa}\left(\frac{kT_v}{I}\right)^2 = 16\ \mu s$$

Once vaporization occurs, a vapor front begins to move into the material, preceded by a liquid front. This situation is depicted in Fig. 12-3. If the material removal process occurs in a relatively short period of time, the energy requirement and pulse time can be reasonably estimated by using the energy balance approach. The energy required to remove a given volume of material is given by Eq. (12-10). For a circular spot of radius a and thickness of material removed z, $V = \pi a^2 z$. It has been assumed that the specific heat and density are constant with respect to temperature and are the same for solid and liquid. The heat of fusion has been neglected. Letting the depth of material removed be a variable and taking the derivative of Eq. (12-10) yield

$$P = \frac{dU}{dt} = \rho\pi a^2 v(CT_v + L_v) \tag{12-18}$$

where $v = dz/dt$ is the speed at which the vapor front moves into the material. The liquid interface between the vapor and solid is extremely thin and is

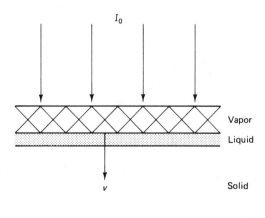

Figure 12-3 Vapor front propagation in a solid.

ignored in this analysis. Because $I = P/\pi a^2$, the speed of the vapor front can be calculated from

$$v = \frac{I}{\rho(cT_v + L_v)} \qquad (12\text{-}19)$$

The time t_p required to remove material to a depth z is then simply

$$t_p = \frac{z}{v} \qquad (12\text{-}20)$$

As an example, consider piercing a 0.5-mm-diameter hole through a 1-mm-thick sheet of steel with $L_v = 6350$ J/g. The energy required, using Eq. (12-14), is

$$U = 11.8 \text{ J}$$

If the irradiance is 10^6 W/cm^2, the pulse length is, by Eqs. (12-19) and (12-20),

$$t_p = 6 \text{ ms}$$

Actual drilling times are 0.1 to 1.0 ms in practice.

If instead of piercing a hole it is assumed that the material is being cut, in the previous example the cutting rate can be estimated from $v = 2a/t_p = 0.05$ cm/6 ms $= 8.3$ cm/s (197 in./min). This is in surprisingly good agreement with experience. Oxygen assist is used to oxidize the metal and to blow away molten material in most metal-cutting applications. In this technique, 60% to 70% of the cutting energy is supplied by the oxidation process and not all the material removed from the kerf is vaporized. The energy balance approach assumes that all the kerf material is vaporized and that no conduction losses occur. These factors are somewhat compensating, which helps to account for the accuracy of the crude energy balance calculation.

When material removal involves pulses with variable power, the energy balance approach is still relatively easy to use if the temporal dependence of the power is known. Using Eq. (12-19), the depth of material removed can be expressed as an integral:

$$z = \frac{1}{\rho(CT_v + L_v)} \int_0^t I(t)\, dt \qquad (12\text{-}21)$$

This integral can be evaluated for a triangular pulse, which approximates a large number of pulsed or Q-switched lasers. Equation (12-21) yields for this case

$$l = \frac{0.5 I_{\max}}{\rho(CT_v + L_v)} \qquad (12\text{-}22)$$

Inherent in this approach is the assumption that the vaporization temperature at the surface is reached in a time much less than t_p and that full penetration is reached before I drops too low to maintain a vapor front.

The rate of material removal for irradiances on the order of 10^8 W/cm^2 becomes limited by the rate at which atoms can leave the surface of the vapor front. The ultimate limiting velocity of atoms in the vapor is the speed of sound in the solid, that is, the maximum speed with which an atom can escape. If the rate of energy input is too great, the vapor does not have time to clear the way for the laser beam, and the vapor becomes superheated; that is, $T > T_v$. Consequently, much of the input energy ends up being absorbed by the vapor. In fact, short, high peak power pulses tend to give rise to a vapor plume that rises above the workpiece and may totally attenuate the laser beam, thereby wasting most of the energy in the pulse. Longer, lower peak power pulses are, in fact, superior for most hole-piercing applications. Such pulses are typically on the order of magnitude of 0.1 to 1.0 ms, and the energy balance approach is reasonable. Because most industrial applications are of this nature, the energy balance approach is the only approach presented here. More rigorous treatments are available in more advanced books [Duley, 1976; Charschan and others, 1977].

12-8 KEYHOLE WELDING

With the advent of CO_2 lasers with 1.0 kW and higher CW power output, it was discovered that the phenomenon of keyholing occurs for continuous seam welding. A similar phenomenon was known to occur for multikilowatt electron beam welding. In this phenomenon a hole is produced in the material, allowing the beam to penetrate relatively unimpeded into the part. Apparently, at power inputs of 1 kW and higher the vapor pressure of the molten material becomes sufficiently high to overcome surface tension and pushes the molten material out of the way, forming a hole, or cavity, that is virtually 100% absorptive. The molten material flows back into the hole after the beam has passed. Generally, the higher the power, the deeper is the cavity that can be formed. This phenomenon is illustrated in Fig. 12-4.

If it were not for the formation of a keyhole, laser (as well as electron beam) welding would be limited to penetration depths of about 1 mm by the thermal properties of metals. Using keyholing, weld depths of several centimeters have been achieved with very high power CW CO_2 lasers.

Swift-Hook and Gick have modeled the keyholing process by use of a linear heat source of total power P extending into the metal a distance a [Swift-Hook and Gick, 1973]. Referring to the geometry depicted in Fig. 12-5, their result for the temperature distribution is

$$T = \frac{P}{2\pi ak} \exp\left(\frac{vx}{2\kappa}\right) K_0\left(\frac{vr}{2\kappa}\right) \tag{12-23}$$

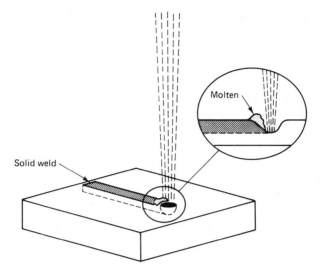

Figure 12-4   Schematic of "keyholing."

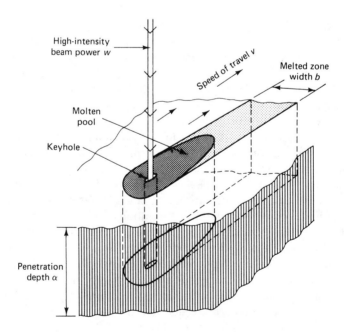

Figure 12-5   Linear heat source model of laser keyholing. (Courtesy of *Optical Engineering* and United Technologies Research Center.)

Here v is the weld speed, K_0 is the Bessel function of order zero, and $r = \sqrt{x^2 + y^2} = x \cos \phi$. The width of the weld can be determined by letting $T = T_m$ and $\phi = 90°$ and solving for $r = w/2$. For high welding speeds, Eq. (12-24) gives a reasonable estimate of weld width.

$$w = 0.484 \frac{\kappa}{v} \frac{P}{akT_m} \tag{12-24}$$

This will apply to most multikilowatt CO_2 laser welding. The graph on page 198 of Duley (1976) can be used to determine whether Eq. (12-24) is applicable.

A second approach based purely on empirical data is given by Banas (1986). By combining data from a number of different sources utilizing many different CO_2 lasers covering a wide range of power levels, a plot of maximum penetration depth d versus power P was generated. The following equation gives a remarkably good fit to the data over the entire power range from about 1 to 20 kW.

$$d = 0.15 \left(\frac{P}{T_m K} \right) \left(\frac{\kappa}{vw} \right)^{1/2} \tag{12-25}$$

In this equation, v is weld speed, T_m is the melting temperature, and w is weld bead width. To apply this equation, we must make reasonable estimates of the average thermal conductivity K, thermal diffusivity κ, and weld bead width.

PROBLEMS

12-1. (a) Using the energy balance approach, estimate the power required to produce a spot weld 3 mm in diameter and 1 mm deep in low-carbon steel with a 1-s pulse from a CO_2 laser.

$$\rho = 7.86 \text{ g/cm}^3, \quad \kappa = 0.07 \text{ cm}^2/\text{s}$$

$$C = 0.7 \text{ J/g K}, \quad k = 0.4 \text{ W/cm K}$$

Estimated values near the melting point are

$$T_m = 1809 \text{ K}$$

$$L_f = 13.8 \text{ kJ/mol}$$

Assume 60% reflectance.

(b) Compare the depth given in part (a) with the thermal penetration depth predicted on the basis of the uniform irradiance model. Remember, at this depth the temperature is only about 35% of the surface temperature.

(c) According to the uniform irradiance model, how long does it take to reach T_m at the surface for the power deduced in part (a)?

12-2. Repeat Problem 12-1(a) using the uniform irradiance model and calculate the temperature at the surface at $t = 1$ s.

12-3. Plexiglas 1.27 cm thick is to be cut by a CO_2 laser at a speed of 5 cm/s. Assuming that the kerf width is 0.25 mm, what is the required power?

12-4. Low-carbon steel of 5-mm thickness is to be cut with a laser at a speed of 1.25 m/min with a 0.5-mm kerf. Using energy balance, estimate the power that must be coupled into the material.

$$k = 0.4 \text{ W/cm K}$$

$$\rho = 7.86 \text{ g/cm}^3$$

$$C = 0.50 \text{ J/g K}$$

$$L_v = 350 \text{ kJ/mol}$$

$$L_f = 13.8 \text{ kJ/mol}$$

$$T_v = 3135 \text{ K}$$

12-5. Mild steel is to be welded with a multikilowatt CO_2 laser. Using the model of Swift-Hook and Gick, estimate the power required to produce a weld of 2-mm width and 4.5-mm depth at 1 m/min.

$$\kappa = 0.09 \text{ cm}^2/\text{s}$$

$$k = 0.4 \text{ W/cm K}$$

$$T_m = 1809 \text{ K}$$

12-6. Estimate the maximum weld depth in mild steel for a 3-kW CO_2 laser welding at a speed of 2 cm/s. Assume the weld bead width is 1.5 mm. (See Problem 12-5 for mild steel properties.)

12-7. A 0.4-mm-diameter copper wire is to be laser welded to the top of a 1-mm-diameter, 70% to 30% brass post. For a focused spot size of 0.4 mm and a melt depth into the post of 0.5 mm, estimate the required energy and pulse length. Assume 50% reflectance. The properties of Cu and brass are as follows:

Cu	Brass (70–30)
$\rho = 8.96 \text{ g/cm}^3$	$\rho = 8.53 \text{ g/cm}^3$
$C = 0.38 \text{ J/g K}$	$C = 0.37 \text{ J/g K}$
$T_m = 1358 \text{ K}$	$T_m = 1230 \text{ K}$
$L_f = 255 \text{ J/g}$	$L_f = 200 \text{ J/g}$

12-8. Holes are to be drilled in 0.5-mm-thick nickel with a diameter of 0.13 mm, using a pulsed Nd–YAG laser with a 4-kW peak power output. Use energy balance to estimate the energy needed and the pulse length, assuming a nearly square pulse. Show that the time to reach vaporization at the surface is negligible, based on the results of the uniform irradiance model.

$$k = 0.91 \text{ W/cm K}$$
$$C = 0.44 \text{ J/g K}$$
$$\rho = 8.9 \text{ g/cm}^3$$
$$L_f = 298 \text{ J/g}$$
$$L_v = 6303 \text{ J/g}$$
$$T_m = 1726 \text{ K}$$
$$T_v = 3187 \text{ K}$$
$$R = 60\%$$

REFERENCES

BASS, M. (ed.), *Laser Materials Processing*. New York: North-Holland, 1983.

BELFORTE, D., AND M. LEVITT (eds.), *The Industrial Laser Handbook*. Tulsa, OK: PennWell Publishing Co., 1986, 69–86.

BERTOLOTTI, M. (ed.), *Physical Processes in Laser–Materials Interactions*. New York: Plenum Press, 1983.

CARSLAW, H. S., AND J. C. JAEGER, *Conduction of Heat in Solids*, 2nd ed. New York: Oxford University Press, 1959.

CHARSCHAN, S. S. (ed.), *Lasers in Industry*. New York: Van Nostrand Reinhold Co., 1972.

———, AND OTHERS (eds.), *The LIA-Material Processing Guide*. Toledo, OH: Laser Institute of America, 1977.

DULEY, W. W., *CO₂ Lasers: Effects and Applications*. New York: Academic Press, Inc., 1976.

———, *Laser Processing and Analysis of Materials*. New York: Plenum Press, 1983.

———, AND OTHERS, "Coupling Coefficient for CW CO₂ Laser Radiation on Stainless Steel," *Optics and Laser Technology*, December 1979, 313–16.

JAHNKE, E., AND F. EMDE, *Tables of Functions*. New York: Dover, 1945.

READY, J. F., *Effects of High-power Laser Radiation*. New York: Academic Press, Inc., 1971.

SWIFT-HOOK, D. T., AND A. E. F. GICK, "Penetration Welding with Lasers," *Welding Research Supplement*, November 1973, 492–99.

Chapter 13

High-Power Laser
Systems and Applications

The purpose of this chapter is to discuss the techniques used in materials processing with lasers and to point out some of the advantages and disadvantages. The primary concern is with heat treatment, welding, cutting, hole piercing, scribing, and marking. The first section deals with beam delivery optics for high-power laser systems. A discussion of alloying and cladding is included. Appropriate theoretical techniques are pointed out and some quantitative information is presented to provide the reader with a feel for laser processing. There is, however, no intent to provide a compilation of laser applications and processing data. Reference may be made to many excellent books of that nature.

13-1 BEAM DELIVERY OPTICS FOR HIGH-POWER SYSTEMS

Materials used for lenses and mirrors in high-power lasers and systems were discussed in Chapter 1. In this section a brief discussion of some for the pertinent ideas concerning Nd–YAG and CO_2 laser delivery systems will be presented.

Nd–YAG delivery systems.   The 1.06-μm wavelength of Nd–YAG lasers does not lend itself to the use of metallic mirrors in beam delivery systems. Consequently, ER dielectric mirrors must be used for that purpose. Since these mirrors will transmit light at other wavelengths, the final mirror may serve as a window for viewing the workpiece, either with a microscope or a closed-circuit TV camera, as shown in Fig. 13-1.

A wide variety of systems is available, some with totally moving optics, others that move only the parts, and hybrids in which the part moves in two or three axes and the beam in two to three axes. In general, moving beam systems are more difficult to align and to maintain. This type of system, however, simplifies part fixturing and may provide greater flexibility and speed.

One advantage that YAG lasers have that CO_2 lasers do not is that the beam can be propagated through optical fibers. A quartz fiber 1.0 mm thick can carry at least 1000 W of average power. The ability to use optical fibers with YAG lasers opens up a wide range of possibilities, such as the following:

1. Splitting the output from one laser between several workstations.
2. Sharing the beam from a single laser between multiple workstations.

Figure 13-1   Viewing workpiece through final mirror by microscopy or CCTV. (Courtesy of Lumonics, Inc.)

The output from the fiber tends to be superior to the input with respect to beam quality for high-power YAG lasers. The output end can be attached to a robot wrist or virtually any type of CNC system for beam manipulation.

CO_2 laser delivery systems.   Because of the high reflectance of metals for the 10.6-μm-wavelength light of CO_2 lasers, metallic mirrors are used to direct the beam in beam delivery systems. Metallic mirrors are also used for focusing, particularly in high-power welding and surface modification applications. Metal mirrors are more rugged than lenses and avoid the problem of thermal lensing. Materials were discussed in Chapter 1.

Similar to YAG systems, there is a wide variety of beam delivery systems using mirrors. There are several types of articulated conduits for use with articulated arm robots. Laser robots have been designed in which the beam is delivered through the robot body to the wrist. A very common laser–robot system is the gantry-robot type. In these systems, typically, the beam is transported in two or three linear axes and two rotational axes. Large parts, such as sheet metal or car bodies, may be moved in one linear axis to accommodate their size and to facilitate part loading and unloading. The z-axis (focusing axis) very often will have an adaptive control for accurately maintaining focus distance and nozzle standoff distance. The sensor is usually capacitive for metallic applications, but optical triangulation (see Chapter 9) is also used. The total travel distance for this redundant axis is around 10 to 15 mm.

Unique optical problems are associated with large CO_2 laser systems. One of these is pointing stability, which refers to the allowable variation in the direction in which the laser beam points as it exits the laser resonator. Typical pointing stabilities are around 0.1 mrad. This means that if the beam direction shifts by 0.1 mrad the point at which the center of the beam strikes a target 10 m from the laser will shift by 1 mm. If a lens or focusing mirror of focal length f is placed in the beam path, the shift in pointing direction θ will cause a lateral shift Δ of the focused spot given by $\Delta = f \tan \theta$, since θ is small. (It is left as a problem to prove this relationship.) Thus, if $\theta = 0.1$ mrad and $f = 0.25$ m, the shift will be 0.025 mm, which is about 0.001 inch. This is very acceptable in most applications. If the pointing direction shifts too much, there is a danger of clipping apertures or running into severe aberration problems with the focusing element.

Other problems that arise relate to the laser optics covered in Chapter 7. Since the laser beam is diverging, unless converged by a telescope, the focused spot size and depth of focus will vary with distance. Depending on the optical design of the laser, this problem may be severe or minimal. A simple way to minimize the problem is to employ a window or output mirror that acts as a converging lens to produce a waist somewhat downstream from the laser. By producing a larger waist and placing it closer to the workpiece, the divergence and therefore the spot-size and depth-of-focus variations are minimized. Problem 13-2 provides an illustration of the effect of different laser resonator designs on beam propagation in a large beam delivery system.

13-2 SURFACE HARDENING

Laser surface hardening (heat treatment) is a process whereby a defocused beam (generally from a 1.0-kW CW or higher-power CO_2 laser) is scanned across a hardenable material to raise the temperature near the surface above the transformation temperature. Normally, the cooling rate due to self-quenching by heat conduction into the bulk material is sufficiently high to guarantee hardening. Steel with over 0.2% carbon and pearlitic cast irons are readily hardened by this technique.

There are several potential advantages to laser surface hardening:

1. The low amount of total energy input to the part produces a minimal amount of distortion, and frequently parts can be used as is after laser hardening.
2. The depth of hardening is relatively shallow; thus desirable properties of the substrate metal, such as toughness, are retained.
3. Selective areas can be hardened without affecting the surrounding area.
4. In some cases, the desired wear characteristics can be achieved in a lower-cost material.

The chief disadvantages to laser hardening are the following:

1. Low-area coverage rate, which limits the process to low production rates or selective hardening.
2. The high reflectance of metals of 10.6-μm radiation means that absorptive coatings must be used.

Hardening with a laser beam does not differ fundamentally from any other hardening process. In steel or iron the temperature at the surface is raised above the martensitic transformation temperature, but is maintained below the melting point, as the beam scans the surface. The cooling rate near the surface can easily exceed 1000°C/s; so the result is a hardened zone down to a depth at which the temperature exceeded the lower martensitic transformation temperature. Figure 13-2 is an illustration of this process. A defocused beam of approximate size d produces a hardened track of approximately the same width and a hardened depth t determined by the laser power and coverage rate.

Most CO_2 laser beams are not suitable for heat treating as they come

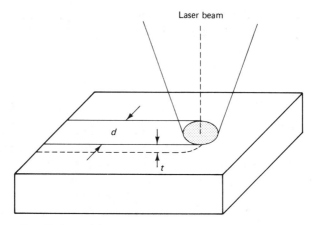

Figure 13-2   Defocused laser beam incident on workpiece for heat treating.

from the laser due to irregularities in the power density (irradiance) in the beam cross section. Gaussian and other lower-order modes are not acceptable because of the highly peaked irradiance distribution even if irregularities are not present. Single higher-order-mode beams are especially suited to hardening because of the large irradiance near the edges of the beam; this factor tends to compensate for lateral heat loss and produces a more nearly rectangular heat-treat-track cross section. Unfortunately, it is difficult to maintain a high-quality, high-order mode over a long period of time. Also, the beam may not be symmetrical, which complicates matters if contour tracks are required.

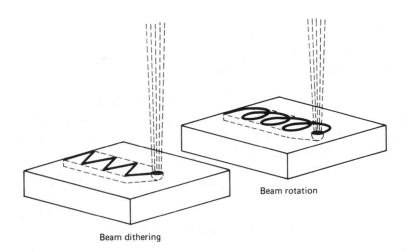

Figure 13-3   Active methods for spreading out a laser beam for surface treatment applications.

Some method to produce a nearly uniform average irradiance profile over a specified spot area is used in many production applications. It may be achieved by rotating the beam optically, thereby producing an overlapping spiral track or by dithering the beam (rocking the lens or a mirror) perpendicular to the track, thus producing a zigzag pattern. Figure 13-3 is an illustration of these techniques. Other techniques, referred to as beam integration, use stationary optical devices to produce relatively uniform beam images or spots. Some are basically hollow, polished copper tubes or pipes. The irregular beam enters one end and a fairly uniform beam exits at the other. Another device, available from Spawr Optical Company, is a segmented mirror. Each of about 32 segments independently forms a rectangular image of the light that strikes it. The superposition of the images formed by the irradiated segments provides a remarkably uniform rectangular spot in the image plane. Spot size can be varied by the use of additional optics. Figure 13-4 is a photograph of a Spawr segmented mirror.

Actual heat-treat-track profiles agree quite well with calculated ones. Numerical integration, finite difference, and finite element analysis techniques have all been successful in predicting heat-treat-track profiles for various irradiance distributions. Depth of penetration can be reasonably well predicted on the basis of the uniform irradiance model, which has an error function solution. The chief difficulty with any theoretical approach is the uncertainty in the value of the absorption coefficient[1] and the thermal parameters of the material.

Figure 13-4 Segmented mirror for beam integration. (Courtesy of Spawr Optical Research, Inc.)

---

[1] Ratio of absorbed power to incident power.

Commonly used absorption coatings are black spray paint, graphite, phosphate, and zinc phosphate. It appears that paint is most effective at the higher power levels (~5 kW) because a fairly thick coating is required. Manganese phosphate (Lubrite) is a common industrial coating and serves as a suitable absorptive coating for lower power levels of around 1 kW. Zinc phosphate has comparable properties to manganese phosphate, and graphite sprays work well but are generally expensive.

The absorption coefficient for uncoated steel or iron may be as low as 10% to 40%, depending on surface condition. A suitable coating raises the absorption coefficient to 60% to 80%.

It is important that the proper thickness of coating be used. If the coating is too thick, too much energy is wasted ablating the coating. If the coating is too thin, it completely burns off before the beam has passed, and bare metal reflectance occurs for part of the pass. What apparently happens when the coating thickness is correct is that heat is absorbed by the coating and transferred to the metal by thermal conduction at a rate that prevents the coating from being totally destroyed. In effect, the metal provides a heat sink for the energy entering the coating, and the coating is removed at a rate that allows maximum coupling of beam energy into the part.

Figure 13-5 contains micrographs of heat treat tracks in 1045 steel and pearlitic cast iron. Hardness in the martensitic region is near the maximum predicted by the carbon content. A thin overtempered zone may exist at the

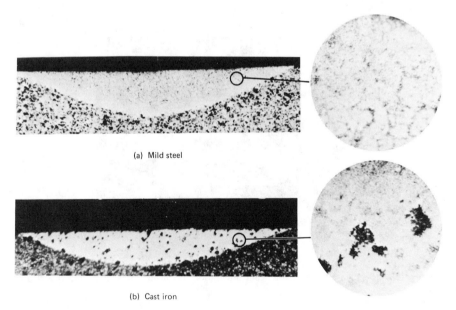

(a) Mild steel

(b) Cast iron

Figure 13-5 Micrographs of heat treat tracks. (Courtesy of Saginaw Steering Gear, Division of GMC.)

(a)   (b)

Figure 13-6 Laser heat-treated parts. [(a) Courtesy of Saginaw Steering Gear, Division of GMC. (b) Courtesy of Electro-Motive Division of GMC.]

border of the hardened area. The ideal cross section for a heat treat track is a rectangle to minimize overlapping, variation in hardened depth, and the amount of overtempered zone at the surface.

Ideally, an inert cover gas should be used during laser heat treatment to minimize oxidation of the metal and flame formation from the coating. It is common practice, however, to use air for economic reasons. A strong cross flow should be used to protect the final optics from smoke and spatter from the coating.

Figure 13-6 contains photographs of production laser heat-treated parts. Figure 13-7 is a photograph of an industrial heat-treatment setup. Figures 13-8 and 13-9 contain plots of penetration depth versus coverage rate for two different materials.

Optimizing penetration depth in terms of coverage rate is a trade-off between rapid and slow heating. If the surface is heated too rapidly, little heat enters the metal before melting occurs, and a shallow depth is obtained. If the coverage rate is too slow, a substantial amount of heat is lost by thermal conduction. Generally, higher power is required to achieve greater hardened depth. To some extent, a narrower track can be run to increase power density, but narrow tracks result in a significant amount of heat being lost by lateral (parallel to the surface) heat conduction. Typical hardened depths in mild steel are 0.5 mm at 1 kW, 1.0 mm at 2.5 kW, and 2.0 mm at 5.0 kW.

Figure 13-7   Industrial laser heat-treating system. (Courtesy of Rofin-Sinar, Inc.)

Hardened depth and coverage rate may be enhanced by using a thin rectangular beam. This reduces the lateral heat loss substantially. Absorption can sometimes be enhanced by altering the angle of incidence from normal. The absorptance of metals is not always maximum for normal incidence. In fact, for a linearly polarized beam, high absorption can be ob-

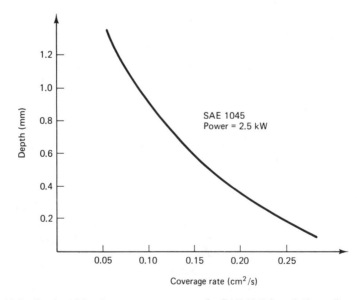

Figure 13-8   Hardened depth versus coverage rate for SAE 1045 for a 0.12-cm diameter beam.

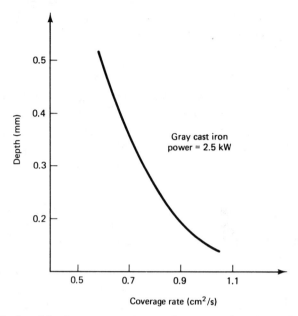

Figure 13-9   Hardened depth versus coverage rate for gray cast iron. Beam diameter, approximately 5 mm.

tained at very large angles when the polarization is parallel to the plane of incidence, due to the fact that the Brewster angle is very large, generally over 70°.

13-3 WELDING

There are basically two types of laser welding, CW and pulsed. Seam welding is done both by CW CO_2 and overlapping pulses with Nd–YAG lasers. Pulsed or spot welding is done chiefly with CO_2, Nd–YAG, Nd–glass, and, to a lesser extent, ruby lasers.

The main advantages of laser welding are as follows:

1. Minimum heat input, which results in very little distortion.
2. Small heat-affected zone (HAZ) because of the rapid cooling.
3. Narrow, generally cosmetically good weld bead.
4. High-strength welds.
5. Easily automated process that can produce very precisely located welds.
6. Weld some metals difficult to weld by other techniques, especially dissimilar metals.

7. Weld in areas difficult to reach with some other techniques.
8. Frequently faster than other techniques.

Disadvantages of lasers in welding are as follows:

1. Extremely hard weld bead in hardenable materials; cold cracking, and hot cracking may be problems due to rapid heating and cooling.
2. High capital investment compared with most techniques, but this factor may be more than offset by improved part quality, high laser up-time (usually over 90%), and relatively low operating and maintenance costs.

Continuous laser welding. At present, continuous seam welding is done only with CO_2 lasers, usually of 500 W or higher power level. Somewhere around 300 to 500 W, keyholing begins to play a role in continuous welding, but melting by thermal conduction is predominant up to around 1.0 kW. At multikilowatt levels, keyholing is the predominant phenomenon.

Theoretical approaches that give reasonable agreement for conduction-limited welding are simple energy balance, the uniform irradiance model, and numerical, finite difference of FEM approaches. The biggest problem with calculations in this regime is the uncertainty in the reflectance of the molten metal, which may approach that of the solid metal.

The method of Swift-Hook and Gick (see Chapter 12) works well for CW welding when keyholing is predominant. It is known that maximum penetration varies approximately as $P^{0.7}$, where P is power. The energy balance approach will give ballpark results if judicious guesses are made regarding conduction and reflection losses.

Nearly all power is absorbed in the keyhole, and very efficient welds can be made at high power levels. Weld efficiency is defined as the ratio of the energy used to melt material to the total energy input. A graphic example has been observed in the welding of the same parts at different power levels. In the welding of low-carbon to mild steel, for instance, both 2.5 and 5 kW were used to achieve approximately 5.5-mm penetration. Parts welded at 2.5 kW could not be handled with bare hands due to heat loss from the weld zone by conduction. Parts welded at 5 kW, at about twice the speed, could be handled without discomfort.

Cover gases play an important role in welding operations, and laser welding is no exception. Some unique problems, however, are associated with cover gases and their method of application in laser welding. Although not unique to CW CO_2 welding, they will be discussed at this point.

The method of application of the cover gas varies with the type of material being welded and the quality of the weld required. Figure 13-10 illustrates a method that uses a trailing cup. Gas is brought directly onto the weld through the nozzle coaxial with the beam to protect the final optical

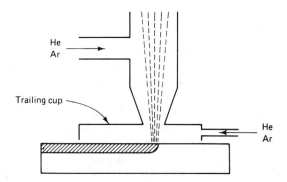

Figure 13-10   Cover gas arrangement using a trailing cup.

element and also is brought into the cup to provide coverage of the bead before and after the beam to prevent oxidation and/or atmospheric contamination. When full penetration is achieved, it may be necessary to provide cover for the underbead as well. Provision of an inert cover gas is especially important for highly oxidizable materials, such as aluminum and titanium.

The most frequently used cover gases are He, Ar, and N_2. Argon, because of its high atomic mass, provides the best coverage or protection against oxidation; however, it interacts with the metal vapor plasma in such a way that more laser radiation is lost in the plasma than with He. Maximum penetration depths are substantially lower for Ar and N_2 than He. Helium has a much higher thermal conductivity than either Ar or N_2 and is able to cool the plasma and reduce the beam interaction. The mechanism for the interaction is free electron absorption, for the photons at 10.6 μm are not sufficiently energetic to cause ionization or atomic excitation. The ionization potential for He is about 1.5 times that of Ar. This factor, coupled with the higher plasma temperature, results in a much greater beam interaction for Ar than He. He can cause excessively rapid cooling in some cases.

It is also likely that the molten pool at the surface is minimized due to the high thermal conductivity of He, thereby allowing better penetration of the beam into the keyhole. It has been observed that in high-speed welding, where maximum penetration is not achieved, deeper penetration is obtained with N_2 than He. It is possible that this situation occurs because the beam is not incident on the plasma as long, and the lower thermal conductivity of N_2 allows more conduction-limited welding, combined with the fact that N_2 is better suited for dispersing the plasma because of its higher molecular mass.

Frequently, the cover gas is applied by means of a tube aimed directly at the weld location if weld bead oxidation is not a serious problem. When welding materials that spatter vigorously, like rimmed steel or high-carbon steel, a strong cross flow of air is required to protect the final optical element.

This cross flow must be placed sufficiently far above the weld area to avoid disturbance of the molten bead and dilution of the cover gas.

Figure 13-11 contains a micrograph of a weld bead cross section. The high aspect ratio and small HAZ are evident. Figure 13-12 is a photograph of two sizes of pipe seam welded by CO_2 laser continuous welding.

Pulse welding.   Numerous materials can be pulse welded with lasers. CO_2 lasers can be used for pulse welding of small parts, called microwelding, but are more frequently applied to spot welding of larger parts. This process can be accomplished by electronic pulsing or simply opening and closing the shutter on CW lasers. Pulse lengths of a few tenths of seconds will yield spot welds several millimeters in diameter and weld depths of a few millimeters for multikilowatt lasers.

Pulsed laser welding is done with Nd–YAG and Nd–glass lasers. Nd–glass lasers are well suited for low-repetition-rate spot welding on small parts, such as wires to terminals and other electronic components.

For high-speed spot welding or seam welding of such items as electronic packages, Nd–YAG lasers are much more suitable. Figure 13-13 is a photograph of a commercial Nd–YAG welding system. Excellent hermetic welds are made by overlapping pulse welding with Nd–YAG lasers.

Figure 13-11   Micrograph of laser weld.

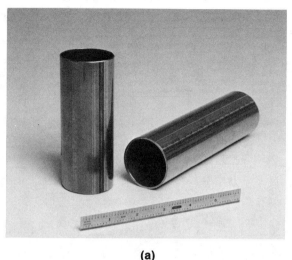

(a)

(b)

Figure 13-12 CO_2 laser-welded pipe. (Courtesy of Rofin-Sinar, Inc.)

Basically, the setup for welding with Nd–YAG or Nd–glass lasers does not differ greatly from CO_2 laser welding. Cover gases should be used whenever contamination or oxidation is a problem and a cross flow may be required to protect the optics.

Calculations for ballpark estimates of pulse times, penetration depth, and/or power requirements can be made by simple energy balance. Typically, pulse lengths are in milliseconds with power densities of 10^5 W/cm^2.

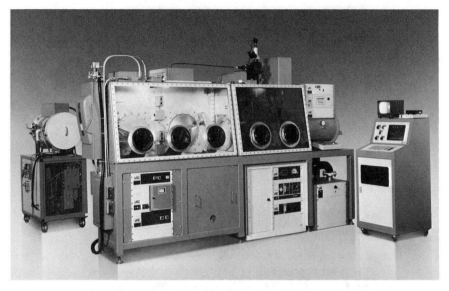

Figure 13-13 Glove box fiber-optic Nd–YAG (laser not shown) welding system with atmosphere-controlled glove box assembly and welding compartments. (Courtesy of Lumonics, Inc.)

13-4 CUTTING

Industrial laser cutting is done with CW or pulsed CO_2 and high-repetition-rate pulsed Nd–YAG lasers. The process is a gas-assist technique in which gas, under pressure, forces molten material from the kerf. Oxygen is used with oxidizable materials to increase cutting speed.

The advantages of laser cutting are essentially the same regardless of the type of laser used and may be the same as for other processing applications. These advantages include the following:

1. Lack of tool contact
2. Ease of automation
3. Small HAZ
4. Narrow, high-precision kerf
5. Frequently higher speed than other methods
6. Ability to reach difficult-to-reach areas, including working through glass
7. Cuts low-machinability materials with relative ease

Some disadvantages are the following:

1. High capital investment. In systems where three-dimensional contouring is required, the system cost may be ten times that of the basic laser.
2. Cut edges may be tapered and serrated when cutting thick sections.
3. The cost of oxygen or other cutting gas is substantial, although not necessarily greater than for other gas-cutting techniques.

Continuous laser cutting.   Generally, laser cutting is a gas-assisted process. It is possible to cut some plastics with the CO_2 laser, such as acrylic, without an assist gas, but it is desirable to use the gas assist to help control the vapors that are produced. There are nearly as many techniques for laser gas-assist cutting as there are users. All techniques, however, involve the delivery of a gas under high pressure to the point of incidence of the laser beam in such a way that the molten material is blown away. Figure 13-14 is a photograph of sheet steel being cut by a CW CO_2 laser. In this case, a coaxial laminar flow nozzle is used, which means that the beam and gas jet are coaxial. The nozzle has a special elliptical cross section leading into a straight throat, which tends to provide a columnar flow of gas on the part without turbulence. The space between the nozzle and the workpiece is about 0.5 to 1.0 mm and the pressure in the nozzle is around 35 kPa. In cutting oxidizable materials, such as steel or titanium, oxygen or air is used to enhance the cutting rate and quality by oxidation. Excessive oxygen,

Figure 13-14 Sheet-metal cutting system. (Courtesy of Coherent Industrial Group.)

however, can cause self-burning, which will result in discouragingly poor cuts.

Some cutting systems use supersonic nozzles that tend to provide a focusing effect for the gas (similar to the exhaust plume from a rocket). Such nozzles are usually off-axis (not coaxial with the beam), although coaxial systems may be employed. Some systems use a window between the focusing optic and the nozzle to keep the high pressure in the supersonic nozzle away from the focusing lens or mirror.

Steel around 5 mm thick is readily cut with CO_2 lasers in the 500-W range with very fine edge quality, and thicknesses of low-carbon steel to 15 mm have been cut at 2.5 kW. Most metals can be cut with CW CO_2 lasers, along with a wide variety of other materials, including composites, glass, quartz, plastics, ceramics, paper, and wood. Surprisingly straight-sided kerfs can be achieved when cutting relatively thick materials as a result of a light-guiding effect due to multiple reflections from the sides of the kerf. Figure 13-15 illustrates this point with 2.5-cm-thick plastic sign letters cut at 1 kW. The same phenomenon occurs to some extent when cutting other materials. The cross section of a kerf typically looks something like the sketch in Figure 13-16.

The amount of undercutting and taper is controlled, to some degree, by the location of the focal point. In metal cutting it is usually best to focus at the surface, but for thick, nonmetallic materials it is frequently beneficial to focus below the surface. In the case of the plastic letters in Fig. 13-15, the

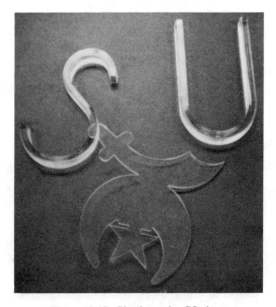

Figure 13-15   Plastic cut by CO_2 laser.

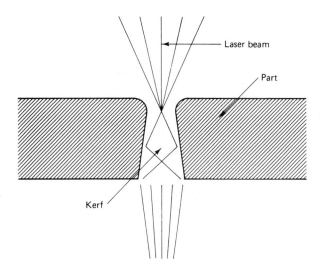

Figure 13-16   Kerf produced by laser cutting.

focal point was approximately one-fourth of the thickness of the part (0.64 cm) below the surface. Figure 13-17 is a photograph of a CO_2 laser-cut saw-blade body.

The importance of maintaining good control of gas pressure and flow rate during cutting cannot be overemphasized. Accurate control of the cut-

Figure 13-17   Laser-cut saw-blade body. (Courtesy of Rofin-Sinar, Inc.)

ting nozzle relative to the work is vital. In the case of large sheet material or parts with z-axis contours, some means of positioning the focusing column, even for very small variations in z axis, is required. In complicated contouring, four- or five-axis numerical control with automatic location sensing and feedback control of the focusing column may be needed. In simple cases where relatively flat stock is to be cut, a two-axis numerical control system with automatic position sensing and feedback control on the z axis may suffice. Feedback control may be accomplished through capacitive coupling to the part, air jet height measurement, and mechanical or optical position sensing. The type of sensor that should be used is a function of the nature of the part's shape, material, and surface condition. A simple technique that is effective on flat material is to float the focusing head mechanically by means of balls that roll on the sheet.

Little, if any, damage to focusing optics due to spatter or smoke occurs during gas-assist cutting. This is true even when making blind cuts (intentionally or otherwise) because the molten material is effectively blown through the kerf or out from under the nozzle in the case of blind cuts. On the other hand, there may be a significant hazard to the operator(s) from vapors or particulates or both given off during cutting, especially of plastics, some of which produce highly toxic substances when heated with a laser.

Pulsed laser cutting.   Basically, there are two reasons for using pulsed laser cutting. There may simply not be sufficient power in low-average-power lasers to melt or vaporize the material without pulsing to achieve high peak power. For example, a Nd–YAG laser with a CW power of only a few watts, when rapidly pulsed, can be used to cut through thick-film resistor material in thick-film circuits to adjust resistance at a rate of several resistors per second. When cutting refractory materials, such as ceramics, it is frequently desirable to use pulsed cutting to minimize the HAZ and thereby reduce microcracking at the cut edge.

A leading application of pulsed laser cutting is the manufacture of aircraft engine parts. The low-machinability, superalloy materials used in aircraft engines, such as Waspalloy and Hastelloy-X, must be cut and drilled by EDM, ECM,[2] or laser. Both turbine blades and combustor parts are being cut and drilled by using CO_2 and Nd–YAG lasers. A laser in combination with a four- to six-axis numerical control system is a powerful and flexible tool for applications involving the cutting of the thousands of holes and slots required in aircraft combustor parts. Only a program correction is required to make a design modification or a new program to process a different part.

Small holes may be cut by a technique referred to as trepanning, which

[2] EDM is electrical discharge machining; ECM is electrochemical machining.

is accomplished by bringing the beam off center through a rotating lens, as indicated in Fig. 13-18.

Effect of polarization on cutting.   Many modern CO_2 lasers produce a linearly polarized output beam. Linear polarization has a deleterious effect on the cutting of ceramics and metals unless they are very thin. If the cutting is done in the direction of the polarization direction, maximum depth, speed, and cut quality are obtained. Under this condition the maximum reflectance occurs from the side walls of the kerf, and maximum absorption occurs on the front wall, where it is desirable. When the direction of polarization is perpendicular to the side wall, maximum absorption occurs there and minimum absorption on the front wall. This causes excessive heating near the top of and on the sidewalls of the kerf. Cut quality is low, as are speed and penetration. The kerf produced is wider than optimum.

When cutting holes or other shapes, the kerf walls become tapered where the direction of polarization makes an angle between zero and 90° with the wall due to the variation in absorption with polarization angle.

A simple test that can be used to determine whether or not a beam is polarized is to cut a circular plug out of 3- or 4-mm-thick low-carbon steel. If the plug falls out or can be easily pushed out, polarization is not a problem. If, on the other hand, the plug cannot be pushed through because of excessive wall taper, some degree of polarization is present in the beam.

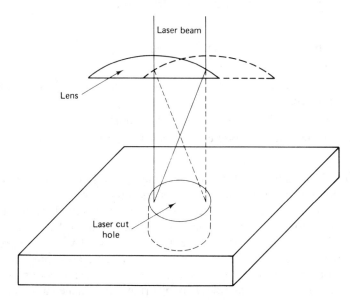

Figure 13-18 Trepanning for cutting small holes.

The cure for a polarized beam is what is generally referred to as an enhanced cutting accessory. This is a set of two to four mirrors upon which are placed birefringent thin films that collectively act as a quarter-waveplate. The direction of polarization must be known, and the device must be accurately aligned relative to that direction to convert the linearly polarized light to circularly polarized light. Circularly polarized light has the same effect on cutting as randomly polarized light; that is, it is directionally insensitive.

To ensure optimum operation of the circular polarizer, the output of the laser must be completely and stably polarized. To guarantee this, some manufacturers place a mirror in the laser cavity with a birefringent coating that acts as a polarizer by selectively absorbing power associated with one direction of polarization more than the orthogonal one. This also fixes the direction of polarization in case it might have a tendency to drift.

13-5 LASER MARKING

Laser marking is a process whereby serial numbers or other identification, including logos, is placed on parts by evaporating a small amount of material with a pulsed CO_2 or Nd–YAG laser.

The advantages of laser marking are as follows:

1. High-speed, easily automated method.
2. No mechanical contact with the part.
3. Can be done through transparent enclosures and in otherwise inaccessible areas.
4. It is trivial to change patterns or index numbers.
5. Can be tied in.with computer inventory control systems.

The major disadvantage is the relatively high capital investment, but it may be offset by improved inventory control and quality control.

A 10-W, average-power Nd–YAG laser Q-switched by an acousto-optical modulator is capable of marking any material not transparent to the radiation. Most modern systems average over 50 W. Even transparent materials, such as acrylic, can be marked if a suitable absorbing material is placed in or on the part. The anodization can be cleanly removed from anodized aluminum, for example, leaving bright, undamaged aluminum, with 8-W average power and a pulse rate of 1000 Hz. The marking or engraving speed is about 5 cm/s with the beam defocused close to 6 mm to widen the lines.

Rapid marking over a field of several centimeters can be accomplished with computer-controlled galvanometric mirrors. Incorporating part movement in the system makes field size essentially unlimited. Figure 13-19 is a photograph of an Nd–YAG laser marking system.

CO_2 lasers can be used for marking many materials transparent to 1.06-μm radiation, and a few watts of average power is all that is required. Pulsing may be used, but it must be realized that electronically pulsing a CW CO_2 laser will produce powers of three to seven or eight times the CW power level. TEA CO_2 lasers produce very high peak powers and are applicable in marking.

One economical, fast, but less versatile technique for marking with pulsed CO_2 lasers is basically a stenciling approach. The pattern to be engraved is defined in a reflective masking material, such as copper, and the entire mask is irradiated and an image is optically produced on the part. If deep engraving is to be done, as in laser wood engraving, the beam can be scanned across a full-size mask in a raster scan pattern.

Excimer lasers are growing in popularity for marking applications because of their unique interaction mechanism with many materials. The excimer laser has the capability to produce extremely fine detail in polymeric materials and glass by use of a mask imaging technique. Figure 13-20 is a photograph of excimer laser engraved parts.

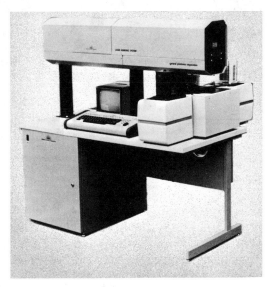

Figure 13-19 Nd–YAG laser marking system. (Courtesy of General Photonics Corp.)

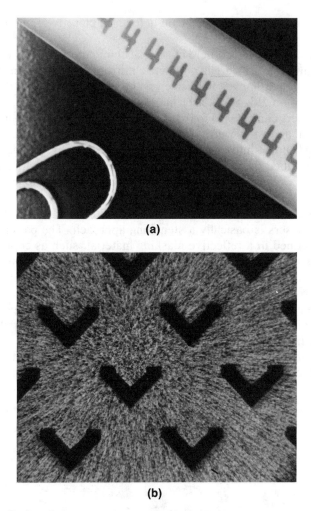

(a)

(b)

Figure 13-20   Excimer laser engraved parts. (a) Wire insulation, induced chemical change; (b) etched marks in integrated circuit. (Courtesy of Lumonics, Inc.)

13-6 HOLE PIERCING

Laser hole piercing, frequently referred to as drilling for short, is chiefly done with pulsed CO_2 and Nd–YAG, although ruby and Nd–glass lasers are sometimes applied. The applications for laser hole piercing are incredibly varied, ranging from piercing holes in cigarette filter paper, baby bottle nipples, and aerosol can nozzles to diamonds and turbine blades for aircraft engines. Hole piercing rates vary from millions of holes per minute for paper to several seconds per hole in turbine blades.

The advantages of hole piercing with lasers are essentially the same as for cutting or welding. It is a technique in which highly reproducible holes can be made rapidly—frequently in materials of low machinability—with no tool contact.

Yet some important disadvantages to laser drilling should be considered. Laser-drilled holes do not have straight, smooth walls. A sketch of a typical laser-drilled hole in a metal is given in Fig. 13-21. In metals, dross may be attached to the underlip of the hole and recast metal may protrude at the top. The material is rarely completely vaporized in metal hole drilling. Instead, some metal is vaporized, and up to 90% is melted and percussively removed from the hole. When a gas assist is used, the molten material is blown from the hole by the gas assist. Molten metal tends to flow up the walls under the influence of the vapor and some resolidifies on the rim of the hole. In addition, molten material may not be completely ejected from the bottom of the hole, thus resulting in dross. The dross is usually easily removed by shot blasting or light grinding, but the recast material on the top is hard and an integral part of the remaining metal. Proper techniques can minimize or eliminate both dross and recast metal on the top surface.

Hole taper is an unavoidable problem associated with beam focusing in fairly high aspect ratio holes. It, too, can be minimized by proper focusing. In multiple-shot drilling, the focal point can be adjusted between shots to control hole shape.

Laser hole piercing in metals has its greatest use in applications in which numerous holes must be precisely located in a highly reproducible manner. The holes routinely placed in turbine blades and turbine combustor parts for airflow control are an excellent example of this type of application, as can be seen in Fig. 13-22.

Estimates of hole-piercing parameters can be made on the basis of simple energy balance and work quite well for vaporization drilling. The pulse time required to drill a given-thickness hole can be estimated by the

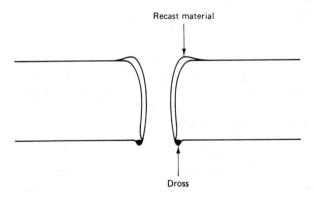

Recast material

Dross

Figure 13-21   Sketch of laser-drilled hole.

(a)   (b)

Figure 13-22   Laser-drilled holes. (Courtesy of Laser Inc.)

technique outlined in Chapter 12 for a propagating vapor front. As noted, however, the development of a metal vapor plume will seriously attenuate the laser power reaching the workpiece. Consequently, pulses must be kept short enough to allow the developing plume to clear the hole. A gas assist or cross flow or both will help reduce this problem. Typical pulse lengths are tenths of milliseconds at power densities on the order tens to hundreds of megawatts.

Holes are routinely laser drilled in ceramic materials. For electrical connections, holes are drilled completely through the boards. Ceramic sheets are scribed by laser drilling a row of blind holes and then mechanically snapping the sheet into two or more parts. Both pulsed CO_2 and Nd–YAG can be used for ceramic drilling, although CO_2 is faster. Pulses should be kept short to maximize vaporization and decrease heat loss by conduction, which produces thermal stresses and hence cracking. Also, rapid pulses minimize the amount of resolidified material in and around the hole, which reduces hole quality. Typically, pulse lengths are one to a few milliseconds with power levels of 50 W for CO_2 lasers and pulse energy of a few joules for Nd–YAG lasers.

The nature of Nd–YAG laser pulses can seriously affect hole quality. Normal relaxation pulses, which consist of a series of initial spikes followed by continuous output, can result in clogged holes due to the material melted by the continuous portion of the pulse. A train of spikes produced by loss modulation provides superior holes in thin materials drilled at relatively low energy per pulse (200 mJ or less per pulse). Proper application of a gas assist also minimizes this problem.

Other materials that are drilled with lasers are rubber and many plastics drilled with CO_2 lasers and diamonds and other gemstones drilled with Nd–YAG lasers.

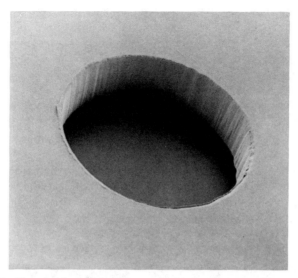

Figure 13-23   Photograph of excimer laser-drilled hole. (Courtesy of Lumonics, Inc.)

Excimer lasers are capable of producing exceptionally clean holes or cuts in polymeric and other organic materials because of the nonthermal interaction mechanism of the uv light and extremely short pulses. Figure 13-23 contains a photograph that illustrates this point.

13-7 ALLOYING AND CLADDING

Two techniques that show great promise for metal surface modification to improve wear, abrasion, corrosion, impact resistance, and strength, while retaining desirable substrate metal properties, are laser alloying and cladding. In both cases, an alloy material, usually a powder, is placed on a metallic surface and melted with a CO_2 laser beam. In alloying, a substantial portion of the substrate material is melted and mixed with the melted powder, thus producing a true alloy layer on the surface after solidification. In cladding, only enough of the substrate is melted to produce a metallurgical bond between the alloy material and the substrate surface. This process is also called *hardfacing*. The technique is similar to heat treatment in that a defocused or integrated CO_2 laser beam is traversed over the region to be melted.

Because the spot size is large (~ 1.0 cm) and melt depths are only on the order of a millimeter, reasonable calculations can be made by using the uniform irradiance model or any of the more sophisticated models used in heat-treatment analysis; however, the heat of fusion should be taken into account.

Some materials used in this work are cobalt- or nickel-based chromium carbides and tungsten carbide nickel alloys. Methods for applying the powders prior to laser melting include powder-solvent slurries, mixing the powder with an organic binder, and plasma spraying. The biggest problem involves applying the powder with the appropriate thickness and applying it consistently. Mechanical scrapers can be used to remove excess powder to a given thickness. An automated spraying technique may also be used. In the most common method the powder is sprayed through a small orifice and simultaneously melted onto the surface by the laser beam as shown in Fig. 13-24.

The advantages of laser cladding and alloying are similar to those found in laser heat treatment and surface melting, with the additional advantages of reduced alloy material consumption and greater control and uniformity of melt depth. The dilution of the cladding by base metal is negligible in

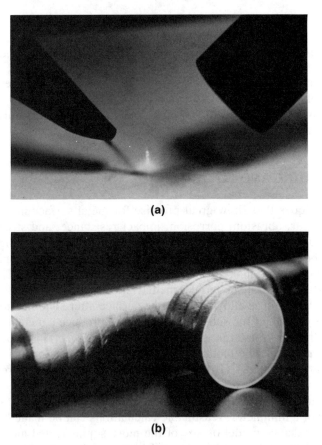

(a)

(b)

Figure 13-24   (a) Laser powder spray cladding simulated with HeNe laser. (b) Laser-clad shaft. (Courtesy of Rofin-Sinar Inc.)

cladding. In alloying, extremely good mixing of the alloying material and substrate metal occurs due to a vigorous mixing action that takes place in the melt pool.

Surface tension and thermal gradient effects produce strong convection currents in the melt puddle. These currents rise at the point of incidence of the beam and flow outward toward the walls and down in a circulatory manner. This mixing causes rippling in the solidified material, both in the direction of the track and transverse to it. Generally, the region of overlap between adjacent tracks will not be flat either. So if a smooth, flat surface is required, a final finishing operation is needed.

13-8 MISCELLANEOUS APPLICATIONS

There are many other material processing applications of lasers. Some notable ones are soldering and desoldering with CO_2 and Nd–YAG lasers, wire insulation stripping with CO_2 lasers, and cutting of silicon for solar cells with Nd–YAG lasers.

PROBLEMS

13-1. If the pointing direction of a CO_2 laser changes by 0.1 mrad, how far off axis will the beam be after traveling 12 m? If there are three right-angle beam benders in the beam's path, does this alter the displacement of the beam? If the beam is focused by a 250-mm focal length lens after traveling 12 m, how far off center is the focused spot?

13-2. For the CO_2 laser resonator designs listed, calculate beam size, focused spot size, focal point location, and depth of focus for a 10% spot size variation at 5 m and 10 m in front of the laser output coupler.
   (a) Resonator design 1: $r_1 = 30$ m, $r_2 = \infty$, $L = 6$ m. Outside of output coupler is flat.
   (b) Resonator design 2: $r_1 = \infty$, $r_2 = 30$ m, $L = 8$ m. Outside of output coupler has a convex radius of 17 m.

13-3. With the aid of sketches, describe the process of laser surface hardening. Explain what the important parameters are and discuss the phenomena of hardening, quenching, and similar processes.

13-4. Describe the process of laser welding. Explain the use of cover gases and other important parameters. Briefly discuss both CW and pulse welding with respect to the type of laser used and the power ranges involved. Explain what keyholing is and what causes it, and point out what power range it occurs in.

13-5. Describe and discuss pulsed and CW laser cutting, and point out what lasers are used for various types of materials.

13-6. Discuss and describe hole piercing with lasers.

13-7. Discuss and describe the processes of alloying and cladding with lasers.

13-8. Discuss the advantages and disadvantages of using lasers in the processes mentioned in Problems 13-3 through 13-7.

13-9. Given a linearly polarized CO_2 laser beam, sketch the effect of cutting with the direction of polarization parallel, perpendicular, and at 45° to the direction of the cut.

13-10. Given that the optical axis of an enhanced cutting accessory is known, show how the direction of polarization of a linearly polarized laser beam should be oriented with respect to that optical axis.

REFERENCES

Banas, C. M., "High Power Laser Welding—1978," *Optical Engineering*, May–June, 1978, 210–19.

Bass, M. (ed.), *Laser Materials Processing*. New York: North-Holland Publishing Co., 1983.

Benedict, G. F., "Production Laser Cutting of Gas Turbine Components," *Society for Manufacturing Engineers Paper No. MR80-851*. Dearborn, MI: The Society, 1980.

Bolin, S. R., "Bright Spot for Pulsed Lasers," *Welding Design and Fabrication*, August 1976, 74–76.

———, "Laser Drilling and Cutting," Society of Manufacturing Engineers Paper No. MR81-365, 1981.

———, "Part 2: Laser Welding, Cutting and Drilling," *Assembly Engineering*, July 1980, 24–27.

Charschan, S. S. (ed.), *Lasers in Industry*. New York: Van Nostrand Reinhold Co., 1972.

Duley, W. W., *CO₂ Lasers: Effects and Applications*. New York: Academic Press, 1976.

Engel, S. L., "Laser Cutting of Thin Materials," Society of Manufacturing Engineers Paper No. MR74-960, 1974.

Engineering Staff of *Coherent Lasers: Operation, Equipment, Application and Design*. New York: McGraw-Hill Book Co., 1980.

Ferris, S. D., H. J. Leamy, and J. M. Poate (eds.), "Laser–Solid Interactions and Laser Processing—1978," *AIP Conference Proceedings No. 50*. New York: American Institute of Physics, 1979.

Gnanamuthu, D. S., "Laser Surface Treatment," *Optical Engineering*, **19**, No. 5, September–October 1980, 783–92.

Jacobson, D. C., and others, "Analysis of Laser Alloyed Surfaces," *IEEE Transactions on Nuclear Science*, **NS-28**, No. 2, April 1981, 1828–30.

Jau, B. M., S. M. Copley, and M. Bass, "Laser Assisted Machining," Society of Manufacturing Engineers Paper No. MR80-846, 1980.

Laser Institute of America, *Use of Lasers in Materials Processing Applications.* Toledo, OH: LIA.

LIA Laser–Material Processing Committee, "Laser Drilling," *Electro-Optical System Design*, July 1977, 44–48.

METZBOWER, E. A. (ed.), *Applications of Lasers in Materials Processing.* Metals Park, OH: American Society for Metals, 1979.

MILLER, F. R., "Advanced Joining Processes," *SAMPE Quarterly*, October 1976, 46–54.

Proceedings of the First International Laser Processing Conference. Toledo, OH: Laser Institute of America, 1981.

READY, J. F. (ed.), "Industrial Applications of High Power Laser Technology," *Proceedings of the Society of Photo-Optical Instrumentation Engineers*, **86**, Bellingham, WA: SPIE, 1977.

———, *Industrial Applications of Lasers.* New York: Academic Press, Inc., 1978.

——— (ed.), "Laser Applications in Materials Processing," *Proceedings of the Society of Photo-Optical Instrumentation Engineers*, **198**, Bellingham, WA, SPIE, 1980.

——— (ed.), *Lasers in Modern Industry.* Dearborn, MI: Society of Manufacturing Engineers, 1979.

REAM, S. L., "High-Power Laser Processing," 1981 SME Heat Treating Conference, MR81-316. Dearborn, MI: Society of Manufacturing Engineers, 1981.

TERRELL, N. E., "Laser Precision Hole Drilling," *Manufacturing Engineering*, May 1982, 76–78.

Index